TECHNOLOGY INVESTMENT:
A GAME THEORETIC REAL OPTIONS APPROACH

T0137803

THEORY AND DECISION LIBRARY

General Editors: W. Leinfellner (*Vienna*) and G. Eberlein (*Munich*)

Series A: Philosophy and Methodology of the Social Sciences

Series B: Mathematical and Statistical Methods

Series C: Game Theory, Mathematical Programming and Operations Research

Series D: System Theory, Knowledge Engineering an Problem Solving

---

SERIES C: GAME THEORY, MATHEMATICAL PROGRAMMING AND OPERATIONS RESEARCH

VOLUME 28

Editor-in Chief: H. Peters (Maastricht University); *Honorary Editor:* S.H. Tijs (Tilburg); *Editorial Board:* E.E.C. van Damme (Tilburg), H. Keiding (Copenhagen), J.-F. Mertens (Louvain-la-Neuve), H. Moulin (Rice University), S. Muto (Tokyo University), T. Parthasarathy (New Delhi), B. Peleg (Jerusalem), T. E. S. Raghavan (Chicago), J. Rosenmüller (Bielefeld), A. Roth (Pittsburgh), D. Schmeidler (Tel-Aviv), R. Selten (Bonn), W. Thomson (Rochester, NY).

Scope: Particular attention is paid in this series to game theory and operations research, their formal aspects and their applications to economic, political and social sciences as well as to sociobiology. It will encourage high standards in the application of game-theoretical methods to individual and social decision making.

The titles published in this series are listed at the end of this volume.

TECHNOLOGY INVESTMENT: A GAME THEORETIC REAL OPTIONS APPROACH

by

KUNO J.M. HUISMAN

Centre for Quantitative Methods CQM B.V.,
Eindhoven, The Netherlands

KLUWER ACADEMIC PUBLISHERS
BOSTON / DORDRECHT / LONDON

A C.I.P. Catalogue record for this book is available from the Library of Congress.

ISBN 978-1-4419-4911-0

Published by Kluwer Academic Publishers,
P.O. Box 17, 3300 AA Dordrecht, The Netherlands.

Sold and distributed in North, Central and South America
by Kluwer Academic Publishers,
101 Philip Drive, Norwell, MA 02061, U.S.A.

In all other countries, sold and distributed
by Kluwer Academic Publishers,
P.O. Box 322, 3300 AH Dordrecht, The Netherlands.

Printed on acid-free paper

All Rights Reserved
© 2002 Kluwer Academic Publishers, Boston
Softcover reprint of the hardcover 1st edition 2002
No part of the material protected by this copyright notice may be reproduced or
utilized in any form or by any means, electronic or mechanical,
including photocopying, recording or by any information storage and
retrieval system, without written permission from the copyright owner.

Contents

Acknowledgments

The research that is presented in this book is the result of a four-year stay at the Department of Econometrics of Tilburg University. I used eight papers as basis for this book. I thank the co-authors of theses papers, Peter Kort (Chapters 2, 3, 4, 5, 6, 7, and 9), Hossein Farzin (Chapter 2), and Martin Nielsen (Chapter 8), for the pleasant and instructive cooperation. I thank Peter Kort and Dolf Talman for proof reading the manuscript of this book.

Last but not least I thank Inge, my and Inge's family, and all my other friends for their understanding and support during the last five years.

Kuno Huisman
July 2001
Best

Acknowledgments

Chapter 1

INTRODUCTION

This chapter is organized as follows. The economic problem on which this book focuses is motivated in Section 1. The two tools used to study this economic problem, which are real options theory and game theory, are discussed in Sections 2 and 3, respectively. Section 4 surveys the contents of this book. In Section 5 some promising extensions of the research presented in this book are listed.

1.   TECHNOLOGY INVESTMENT

Investment expenditures of companies govern economic growth. Especially investments in new and more efficient technologies are an important determinant. In particular, in the last two decades an increasing part of the investment expenditures concerns investments in information and communication technology. Kriebel, 1989 notes that (already) in 1989 roughly 50 percent of new corporate capital expenditures by major United States companies was in information and communication technology. Due to the rapid progress in these technologies, the technology investment decision of the individual firm has become a very complex matter. As an example of the very high pace of technological improvement consider the market for personal computers. IBM introduced its Pentium personal computers in the early 1990s at the same price at which it introduced its 80286 personal computers in the 1980s. Therefore it took less than a decade to improve on the order of twenty times in terms of both speed and memory capacities, without increasing the cost (Yorukoglu, 1998).

In the beginning of the twentieth century technological developments did not show such a rapid progress compared to recent years. Therefore, the technology investment problem of the firm mainly was a timing

1

problem, in which the optimal time to replace the current technology had to be determined. For example, one of the technology investment decisions of a railway company dealt with the decision when to replace its steam shunters with diesel shunters. Up to the present day most railway companies still work with diesel shunters.

Nowadays, a firm should take into account that the current state of the art in information and communication technology will be old fashioned in a few years. Thus the investment decision problem is no longer only a question of when to adopt a new technology but also a question of which technology should be adopted. Therefore, in order to design a theoretical framework that is useful to analyze the technology investment decision, it is important to consider models in which several new technologies appear. The timing of the technology investment is (still) very relevant. The reason is that due to the rapid technological progress of information and communication products the prices of those products drop significantly over time. As an example in Figure 1.1 the price development is drawn of two Intel Pentium III processors within the Netherlands.

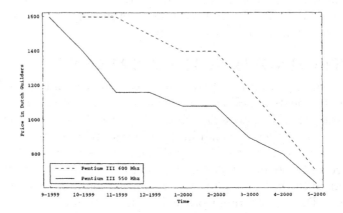

Figure 1.1.   Price development of two Pentium III processors within the Netherlands (source: Personal Computer Magazine).

Another significant feature of the last decade is that firms more and more face competition on their output markets. One reason is the abolition of monopolistic markets created by government. In the Netherlands examples are the opening of the markets for telecommunication, railway, and power supply. Until September 1995 KPN Telecom was the only provider on the mobile telecommunication market in the Netherlands. Due to the legislation of the European Community concerning the liberalization of telecommunication markets, the Dutch government organized a contest with as prize a license to operate a mobile telephone company. Libertel won that contest and its network came into use in September 1995. Telfort entered the market in September 1998 and after that Ben and Dutchtone started their services at the end of 1998. Currently these five players are still active on this market. It is obvious that due to the entrance of these four rivals, KPN Telecom had to change its investment strategy dramatically.

Another reason for the existence of and development towards oligopolistic markets is the, still ongoing, process of mergers and takeovers, which due to legislation will not end up with only one supplier in a market. There are plenty of examples of mergers and takeovers in the last decade. In the car industry we have the merger of Daimler and Chrysler in 1998. In the telecommunications market examples are the takeover of AirTouch by Vodafone (partly owner of Libertel) in 1999 and this year's offer of Vodafone to the shareholders of Mannesmann, who entitled it a hostile takeover. However, when Vodafone succeeds in taking over Mannesmann, it has to hive off Orange (which is owned by Mannesmann), otherwise Vodafone's market share in the United Kingdom would become too large. The result of Vodafone's announcement is that there are already four potential buyers for Orange: France Telecom, KPN Telecom, NTT DoCoMo, and MCI Worldcom. This year's announcement of the merger of the Deutsche Bank and the Dresdner Bank is an example in the financial market. The lawsuit between Microsoft and the government of the United States of America is an example of governmental interference to try to reduce Microsoft's (supposed) monopoly power. The overall result of these mergers, takeovers, and governmental interference is that markets with only one supplier and markets with many suppliers seem to disappear in the long run. Thus, in their private investment decisions, more and more firms should take into account the investment behavior by its competitors nowadays.

The existing literature dealing with the technology investment decision of a single firm can be split up into two categories. The models that belong to the first category, which is called *decision theoretic models*, deal with the technology investment decision of one firm in isolation. On the other hand, in the *game theoretic models* the optimal technology investment strategy of the firm is derived while taking explicitly into account the technology investment actions of the firm's rival(s). From the economic observations described above, it can be concluded that there is a strong need to use game theory with respect to the theoretical modelling of technology investment by the individual firm. A literature overview of the decision theoretic models and the game theoretic models are given in the first sections of Chapters 2 and 4, respectively. Part I of this book deals with decision theoretic models and in Parts II and III two different game theoretic models are considered.

In the following section we discuss the topic of investment under uncertainty in more detail, whereas in Section 3 we concentrate on investment under competition.

2.   INVESTMENT UNDER UNCERTAINTY

Investment is defined as the act of incurring an immediate cost in the expectation of future rewards (see p. 3 of Dixit and Pindyck, 1996). Most investment projects possess the following three characteristics: irreversibility, uncertainty, and possibility of delay.

An investment is irreversible when the investment cost is a sunk cost, that is it is impossible to recover the investment cost once the investment is made. This surely holds for investments in information and communication goods. It is impossible to sell a one-year-old personal computer for the same price as for which it was bought. More generally, most industry or firm specific investments are irreversible. For example, the marketing and advertisement expenditures of KPN Telecom are firm specific and cannot be recovered, i.e. KPN Telecom cannot sell this investment project to another telecommunication company. An example of an industry specific investment is the construction of a new mobile network by Libertel. This investment will be (at least partially) sunk, because whenever it is no longer profitable for Libertel to exploit this network it will not be profitable for another mobile provider as well. Due to the lemon's problem (see Akerlof, 1970) a lot of investments that are not firm or industry specific are also irreversible.

An investment project (almost) always has to deal with uncertainty. For most investments the future revenues are stochastic, due to uncertainties in, e.g., the firm's market share and the market price. It is also possible that the investment cost is uncertain, which is the case in many infrastructure projects. For example the actual costs of the *Oosterschelde Stormvloedkering*, which is one of the last projects of the Dutch Delta works, turned out to be much higher than what was forecasted.

From a technical point of view it is almost always possible to defer an investment for some time. This possibility to delay an investment gives a firm flexibility. Though, economically this postponement can be costly in the sense that the firm looses market share if it refrains from the investment. On the other hand, by postponing an investment the firm can acquire more information about the investment project, for example concerning the market conditions.

The net present value method is the commonly used (and taught) method to evaluate investment projects (see for example Brealey and Myers, 1991). This method states that an investment should be undertaken when the expected (discounted) present value of the revenue stream resulting from this investment project exceeds the expected present value of the expenditures. However, the underlying assumption of the net present value method contradicts the characteristics of investments we mentioned before. More precisely, the net present value method assumes that either an investment project is reversible or when it is irreversible it is a now or never decision, that is, the firm can undertake the investment project today or never. As a result, applying the net present value method leads to suboptimal investment decisions. Especially the ignorance of the possibility to delay is an important abuse, since most investment projects are irreversible. The real options literature succeeds in explicitly valuing this so-called option value of waiting.

In the real options theory the analogy between a firm's investment opportunity and a financial call option is exploited. A financial call option gives the holder the right, but not the obligation, to buy one piece of the underlying derivative (e.g. stock, bond) for a specified price (before or) at a specified time. See Hull, 1993 and Merton, 1992 for a detailed exposition of the financial options theory. Similar to a financial call option, an investment opportunity gives a firm the right, but not the obligation, to carry out some investment project. Note that, following the analogy, an investment project is an infinitely lived call option on a dividend paying derivative. From the financial options theory it is known that such an option should only be exercised when the option is sufficiently deep in the money, that is when the current price of the underlying derivative is sufficiently larger than the exercise price. Therefore an investment project should only be undertaken when the net present value exceeds the option value of waiting.

Due to the close link with financial options theory, most real options models assume that the revenue stream of the investment project follows some geometric Brownian motion process. A geometric Brownian motion process is a continuous time stochastic process of which the increments are distributed according to a normal distribution. In McDonald and Siegel, 1986 the basic continuous time real options model is examined. In that model a firm can acquire a project, of which the value follows a geometric Brownian motion process, by making an irreversible investment. Mcdonald and Siegel derive an explicit expression for the option value of waiting, and show that for reasonable parameters the optimal investment trigger is twice as large as the net present value trigger, i.e. for an investment to be optimal the value of the project should be twice as large compared to the required value under the net present value method. The basic real options model has been extended in various ways. An excellent overview is given in Dixit and Pindyck, 1996. In Trigeorgis, 1995, Trigeorgis, 1996, Smit, 1997, Pennings, 1998, Lander and Pinches, 1998, and Amram and Kulatilaka, 1998 practical applications of the real options theory are presented and discussed.

However, most of the extensions of the basic real options model assume that the firm is the only one having the investment opportunity, that is, strategic interactions are ignored. In the previous section we stressed the importance of taking strategic interactions into account. In the next section we give an overview of the models that do incorporate these strategic interactions.

3.   INVESTMENT UNDER COMPETITION

Investment models that incorporate strategic interactions make use of game theory. In most of these models non-cooperative game theory is used since in general the firms are competing against each other and there is no willingness to cooperate. For a rigorous introduction to game theory we refer to Fudenberg and Tirole, 1991. In Tirole, 1988 a nice overview of industrial organization models is given. In this book we mainly use and extend the game theoretic concepts presented in Fudenberg and Tirole, 1985, in which a deterministic investment model of two competing firms is extensively analyzed, especially from a mathematical point of view.

One of the first real options models that incorporates strategic interactions is the duopoly model in Smets, 1991. This model is also considered in Chapter 9 of Dixit and Pindyck, 1996 and in Nielsen, 1999. As in the basic real options model the revenue stream of the investment project follows a geometric Brownian motion. However, in this model the actual revenue stream of one firm depends on the investment decision of the other firm. Nielsen proves that due to the introduction of a second identical firm, the first investment will be made sooner. Like in Smets, 1991, also in Pennings and Sleuwaegen, 1998 a model of foreign direct investment in a real options setting is studied.

Other continuous time real options models with strategic interactions are studied in Grenadier, 1996, Baldursson, 1998, Lambrecht and Perraudin, 1999, and Weeds, 1999. Grenadier models the real estate development. In Baldursson's model the firms can adjust their capacity continuously over time and optimal strategies to do so are derived. Lambrecht and Perraudin consider a model with incomplete information, in which they assume that the other firm's profitability of the investment project is not known and there is only one firm that can implement the project. Finally, Weeds models a research and development race between two firms.

Also related are the models in Smit, 1996, Smit and Ankum, 1993, Kulatilaka and Perotti, 1998, and Somma, 1999. The difference with the contributions mentioned above is that, instead of a continuous time model where uncertainty is incorporated by, e.g., a geometric Brownian motion process, in each of these models binomial trees are used to model strategic interactions between firms. Consequently, most of these models are in discrete time.

4. OVERVIEW

This book is divided into three parts. The first part consists of two decision theoretic models in a real options setting. In the second part three game theoretic technology adoption models are considered, whereas in the third part three general game theoretic real options models are presented.

4.1 DECISION THEORETIC MODELS

Chapter 2 starts with a literature overview of the decision theoretic models of technology adoption. The model of Chapter 2, which is a generalization of Farzin et al., 1998, studies the technology adoption of a single firm. A firm can adopt a better technology by making an irreversible investment. The investment cost of a certain technology is assumed to be constant over time. With a better technology the firm can produce more efficiently and will therefore make higher profits. New technologies arrive over time according to a stochastic process and the efficiency increment of a new technology is also stochastic. First we solve the model for the case that the firm may invest only once and after that the multiple switch case is discussed. The multiple switch case is only solvable when the efficiency increments of the new technologies are known beforehand. Finally, the optimal investment strategy is compared with the strategy resulting from the net present value method. It turns out that in making the investment decision the option value of waiting cannot be ignored, which is also shown in a numerical example. In the Appendix to Chapter 2 we give an introduction to a technique that is frequently used in this book: optimal stopping.

The model of Chapter 2 is extended in Chapter 3 by making the investment cost decreasing over time. The motivation for this model feature is illustrated in Figure 1.1. The efficiencies of the new technologies are assumed to be known beforehand. The optimal investment strategy for the single switch case is derived and compared with its net present value counterpart. After that, it is explained why it is not possible to solve this model analytically for the multiple switch case.

4.2   GAME THEORETIC ADOPTION MODELS

In Chapter 4 we first give a literature overview of game theoretic models of technology investment. After that we analyze the most important and basic deterministic model in this research area (see Reinganum, 1981 and Fudenberg and Tirole, 1985) in detail. Two identical firms that are currently active on an output market can make an irreversible investment that will increase their own and decrease their rival's profit. The investment cost is decreasing over time. This investment game is solved using timing games. In the Appendix to Chapter 4 we present an introduction to this specific class of games. The player that moves first in a timing game is called the leader and the other is the follower. Reinganum assumes in her analysis that one of the firms is given the leader role beforehand. In Fudenberg and Tirole's analysis the firm roles are determined endogenously, that is both firms can become the leader by making the investment before its rival. It turns out that the result is that each firm's payoff is equal in equilibrium. The remainder of Chapter 4 deals with the extension of Stenbacka and Tombak, 1994 to the Reinganum-Fudenberg-Tirole model, which is also is considered in Huisman and Kort, 1998a and in Götz, 2000. Stenbacka and Tombak assume that the time between the adoption and successful implementation of the new technology is stochastic.

Chapter 5 is based on Huisman and Kort, 1998b. The model of that chapter is an extension of the basic model by incorporating two new technologies and upgrading. At the beginning of the game none of the two firms is active on the output market. To become active a firm has to pay a sunk cost, for which the firm receives the current best technology. The firm can also decide to postpone the entrance and buy the better technology that becomes available at a known point of time in the future. Furthermore, there is a possibility to upgrade the current technology with the new technology. There are learning effects involved in this upgrading strategy, since it is cheaper to buy the new technology when the firm already produces with the current technology. Two of the nine possible scenarios are worked out in detail.

Chapter 6, which is based on Huisman and Kort, 2001, extends the model of Chapter 4 by adding uncertainty to the arrival process and by considering multiple new technologies. Further it is assumed that the two firms can invest only once. After introducing a new concept in timing games, namely the waiting curve, a general algorithm for solving this kind of technology investment games is presented. The algorithm is clarified by applying it to a specific example.

4.3 GAME THEORETIC REAL OPTIONS MODELS

In Chapter 7 the model of Huisman and Kort, 1999 is presented. This model differs from the model in Smets, 1991 since in that chapter the firms are already active on the output market. We show that abolishing the new market assumption considerably changes the result of Nielsen, 1999. No longer it is always the case that the introduction of a new firm precipitates investment. Furthermore, the model is the stochastic variant of Fudenberg and Tirole, 1985 and we discuss what the effect of the introduction of the uncertainty is on the equilibrium outcome.

The new market model of Nielsen, 1999 is studied in Chapter 8, but then with asymmetric firms. The asymmetry is modelled via the investment costs of the firms. Both the case of negative and positive externalities are considered. In case of negative externalities a firm earns the highest profits when its rival is not active, whereas in the positive externalities case the firm's profit is higher when the other firm is also active, e.g. when there are network effects or when the firms are producing complementary goods. We show that also in this asymmetric setting the introduction of a second firm precipitates investment. The chapter is based on Huisman and Nielsen, 2001.

Chapter 9, which is based on Huisman and Kort, 2000, brings together Chapters 2, 5, and 7. Two firms can become active on a output market, i.e. a new market model is considered, by buying a technology. In the beginning only one technology is available, but at an unknown point of time in the future a better technology becomes available. Both firms can invest only once and the investment cost is constant over time. We show how the equilibrium outcome depends on the expected speed of the arrival of the new technology.

5. EXTENSIONS

In this section we present some promising extensions of the research of this book.

5.1 INCOMPLETE INFORMATION

Thijssen et al., 2001b consider a monopolistic firm that can enter a new market by incurring an irreversible investment cost. It is assumed that the market can either be good or bad. When the market is good it is profitable to start producing and when the market is bad the firm will never produce. In the beginning of the game the firm believes that the market is good with some positive probability. At stochastic points in time the firm receives signals, being either good or bad, about the condition of the market. At these points in time the firm updates its belief that the market is good in a Bayesian way. The probabilities of receiving a good or a bad signal are such that the uncertainty about the state of the market vanishes in the long run. As such the model is a promising extension of the existing real options models, since in the latter models uncertainty never disappears. An interesting extension of this model is to include strategic interactions by the introduction of a second firm (see Thijssen et al., 2001a). The firm that invests first and finds out that the market is good becomes Stackelberg leader and the other firm the Stackelberg follower. The advantage of the other firm is that it leaves the risk of investing in a bad market to its competitor, since after one investment the true status of the market is revealed. In this way there are incentives to become both the first investor and the second investor.

Pawlina and Kort, 2001 extends the basic real options model by making the investment cost subject to a possible upward switch. The value of the project again follows a geometric Brownian motion process. The firm does not know when the switch in the investment cost takes place, but has a certain belief about it. As such the model is also an extension of Lambrecht and Perraudin, 1999. Lambrecht and Perraudin assume that once a firm has invested the other firm is left with a zero payoff. Here the investment cost switch can be interpreted as a reduced profitablity resulting from the investment of the other firm, so that there is still a positive payoff possible for the firm.

5.2   RISK AVERSION

Most real options models assume that either the firm is risk-neutral or that the markets are complete, so that any risk in an investment project can be hedged and as a result the risk free interest rate can be used. In other models (see for example Sarkar, 2000) risk aversion is modelled by adding a risk premium to the discount rate. In van den Goorbergh et al., 2001 the real options literature is extended by considering a risk averse firm in a world with incomplete markets. In the model risk aversion is taken into account by making the firm's utility function concave. First, van den Goorbergh et al. derive the properties that an utility function should possess. After that the existence of a threshold is proved and a specific utility function is chosen for which a comparative statics analysis is carried out.

I
DECISION THEORETIC MODELS

1

DECISION THEORETIC MODELS

Chapter 2

CONSTANT INVESTMENT COST

1. INTRODUCTION

The literature on technology adoption can be divided into two classes: the first class is called decision theoretic models and the second class game theoretic models. See Bridges et al., 1991 for an overview of literature from both classes. In a decision theoretic model the profit of the firm is only influenced by its own technology adoption decisions, whereas in a game theoretic model the profit of the firm is also influenced by the decisions of its rivals. In the second and third chapter we study decision theoretic models of technology adoption. This implies that either the firm is a monopolist or a price taker on its output market. From Chapter 4 onwards strategic interactions are incorporated in the technology investment problem.

Though Baldwin, 1982 does not explicitly model technology adoption, in that paper sequential investments can be looked upon as investments that upgrade the firm's technology in use, while the state variable resembles the efficiency of the best technology that is available. Due to the fact that the investments are not completely reversible, the firm will only carry out those investment opportunities that yield a net present value that exceeds a certain threshold. Of course, this threshold is the option value of waiting.

Nair, 1995 uses a dynamic programming framework to solve the technology adoption problem for a firm. First the model is solved for a finite planning horizon before which a fixed number of new technologies will arrive. Then it is shown that the results can be extended to a model with an infinite planning horizon by using forecast horizon procedures. A time τ is a forecast horizon when the initial period decision is optimal

13

for all models with planning horizon larger than τ. The investment cost and revenue function of a technology may depend on time, which is discrete in the model of Nair. The drawback of the analysis is that there are no expressions for the option value of waiting.

In Rajagopalan, 1999 three technologies are taken into account. The first technology is the one that the firm uses at the beginning of the (infinite) planning horizon, the second technology is the best technology available at that point of time, and the third technology becomes available at a yet unknown point of time in the future. The reason for considering only three technologies is that the author focuses on the impact of investment costs and other factors on the replacement time. Once the firm has adopted a newer technology it can produce against lower marginal costs. Rajagopalan derives how the optimal adoption time of the second technology depends on the probability distribution of the arrival of the third technology. He specifies for what parameters it is optimal: (i) to adopt the second technology immediately, (ii) to wait with adoption for some finite time, and (iii) never to adopt the second technology.

Balcer and Lippman, 1984 model the arrival time and the efficiency of new technologies in a two step procedure. Every time the discovery potential changes, a new technology is invented. The time between two consecutive changes of the discovery potential and the increase of the discovery potential are both stochastic and depend on the current discovery potential. The efficiency of a new technology is a stochastic function of the new discovery potential and the efficiency of the current best technology. To be able to solve the model, Balcer and Lippman assume that the firm's profit increase and the investment cost are both linear functions of the efficiency of the new technology. They show that the firm is going to upgrade its current technology with the current best technology if the technological lag exceeds a certain threshold. Further, if the arrival of a new technology takes too long, adopting an already existing technology may become optimal.

In the models of Baldwin, 1982, Nair, 1995, Rajagopalan, 1999, and Balcer and Lippman, 1984, a general distribution is used to model the arrival of new technologies. We use a Poisson process to model the technology arrivals. This implies that the interarrival times are exponentially distributed, i.e. the time elapsed since the last technology arrival does not influence the probability of a new technology arrival. This will hold especially when the firm does not have a clue of what is going on at the research companies.

We take the real options approach (see Dixit and Pindyck, 1996 for an excellent survey) to be able to explicitly derive the option value of waiting that is present in technology investments. Purvis et al., 1995 and Dosi and Moretto, 1997 also use the real options approach to model a technology investment problem, but both models consider only one new technology. Purvis et al., 1995 try to derive ex ante the effects of uncertainty and irreversibility on the investment decision of Texas farmers, who can change from conventional open lot to free stall housing for their dairy. Dosi and Moretto, 1997 show that lowering the uncertainty about future profits is a better policy to stimulate investment than lowering the investment cost.

Another model that uses the real options approach and incorporates technical change is introduced in Ekboir, 1997. He studies the effect of partial irreversibility and technological change on the purchase and sale of capital. Ekboir shows that if the desired capital stock is in between an upper and a lower bound it is optimal for the firm not to invest. After hitting the upper (lower) bound the firm purchases (sells) new capital and due to technical change the average capital productivity increases (decreases). Therefore both the upper and lower bounds increase (decrease) significantly and the desired capital level increases (decreases). It is possible that the firm is going to buy (sell) more new capital, because the desired capital level is not in between the new bounds. This process stops when the desired capital level is again between the bounds.

The analysis of this chapter adds a multiple switching technology adoption model to the existing real options literature. On the other hand we extend the traditional decision theoretic models on technology adoption with a model in which the technologies arrive according to a Poisson process.

In Section 2 the model is described. This model is a generalization of the model that was introduced in Farzin et al., 1998. The original model of Farzin et al. is used as an example. The single switch case is solved in Section 3 and the multiple switch case is treated in Section 4. In that last section we also point out and correct a mistake in Farzin et al., 1998 (see also Doraszelski, 2001). In Section 5 the net present value method is used to solve the technology investment problem and the outcome is compared with the results of Sections 2 and 3. The last section concludes.

2.   THE MODEL

A risk-neutral firm is considered, whose profit flow is only determined by its own technology choice. The efficiency of a technology is completely captured in one parameter, in such a way that a higher value of the parameter implies a more efficient technology. We use two symbols to refer to this technology efficiency and the technology itself. The efficiency of the technology that the firm uses at time t, $t \geq 0$, is denoted by $\zeta(t)$. With $\theta(t)$ we refer to the efficiency of the most efficient available technology at time t, $t \geq 0$. Of course it must hold that $0 \leq \zeta(t) \leq \theta(t)$, for any $t \geq 0$. The firm's profit flow when the firm produces with technology ζ, $\zeta \geq 0$, equals $\pi(\zeta)$, where $\pi : \mathbf{R}_+ \rightarrow \mathbf{R}$ is an increasing function of ζ.

EXAMPLE 2.1 *Let us analyze the firm that is considered in Farzin et al., 1998. The firm's production function is given by*

$$h(v, \zeta) = \zeta v^a, \qquad (2.1)$$

where $v\,(\geq 0)$ is a variable input, $\zeta\,(\geq 0)$ is the efficiency parameter, and $a \in (0, 1)$ is the constant output elasticity. Further we assume that the output price and the input price are fixed and equal to p and w, respectively. The profit flow equals

$$\pi(\zeta) = \max_{v}\left(p\zeta v^a - wv\right). \qquad (2.2)$$

Solving equation (2.2) yields the following expression for the profit flow of the firm

$$\pi(\zeta) = \varphi \zeta^b, \qquad (2.3)$$

with

$$\varphi = (1 - a)\left(\frac{a}{w}\right)^{\frac{a}{1-a}} p^{\frac{1}{1-a}}, \qquad (2.4)$$

$$b = \frac{1}{1 - a}. \qquad (2.5)$$

Over an infinite planning horizon the firm maximizes its value and discounts with rate $r\,(> 0)$. At the beginning of the planning horizon ($t = 0$) the firm produces with a technology whose efficiency equals $\zeta_0\,(\geq 0)$. As time passes new and more efficient technologies are invented. The firm can not influence the innovation process, i.e. it is assumed to be exogenous to the firm. We assume that $\theta(t)$ follows the following Poisson jump process:

$$d\theta(t) = \begin{cases} u & \text{with probability } \lambda dt, \\ 0 & \text{with probability } 1 - \lambda dt, \end{cases} \qquad (2.6)$$

$$\theta(0) = \theta_0, \qquad (2.7)$$

where $\theta_0 \geq \zeta_0$. Concerning the size of the jump, u, two cases are considered. In the first case u is constant and in the second case u is stochastic.

Let us introduce some more notation. Take $i \in \mathbb{N}$. The efficiency and the arrival time of the i-th technology are denoted by θ_i and T_i, respectively. We use τ_i to refer to the time between the arrival of technologies $i-1$ and i, i.e., $\tau_i = T_i - T_{i-1}$. The jump in the θ process at time T_i is denoted by u_i, thus $u_i = \theta_i - \theta_{i-1}$. We set $T_0 = 0$ and $u_0 = 0$. $N(t)$ is the stochastic variable that counts the number of technology arrivals on the interval $[0, t)$. Therefore,

$$\theta(t) = \theta_0 + \sum_{n=0}^{N(t)} u_n. \tag{2.8}$$

Since θ follows a Poisson process with parameter λ, $N(t)$ is distributed according to a Poisson distribution with parameter λt. In Figure 2.1 two sample paths for the technology process are plotted. In the left panel the jump size is constant and in the right panel the jump size is uniformly distributed.

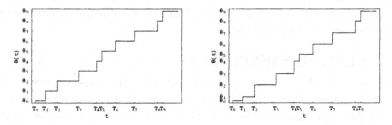

Figure 2.1. Sample paths of $\theta(t)$. In the left panel the jump size is fixed and equal to u and in the right panel the jump size is uniformly distributed on $[0, 2u]$. Parameter values used: $\lambda = 1$, $u = 0.1$, $\theta_0 = 1$.

The firm can adopt a new technology by paying a sunk cost $I \, (> 0)$. In this chapter the investment cost is assumed to be constant over time, whereas in the next chapter the investment cost decreases over time. Due to the constant investment costs, the exponential interarrival times, and due to discounting, the firm will adopt a technology only at its arrival date.

The problem we want to address concerns the timing of the firm's technology switches. Two extreme solutions are easy to identify. The first is to adopt a new technology every time that one becomes available. The advantage of this strategy is that the firm always produces with the most efficient technology. The drawback of this strategy is, of course, the large amount of investment costs that has to be paid. The other extreme solution is to never adopt a new technology. The advantage is that the firm does not have to pay any sunk costs. Since the firm keeps on producing with an (in the long term) inefficient technology, the drawback is that the firm misses the potential higher profits it could have made when a more efficient technology was adopted. The optimal solution will be somewhere in between these two extreme solutions.

The problem that the firm faces is an optimal stopping problem. For an introduction to optimal stopping problems we refer to Appendix A. In our model stopping means that the firm invests and thus adopts a new technology and continuation resembles waiting with investing. Therefore, it is obvious that stopping will be optimal for θ large enough and waiting is optimal for θ low enough. Hence, intuition suggests that there is some unique value of θ for which the firm is indifferent between investing and waiting. That specific θ is denoted by θ^* and is called the critical level or threshold (value). Once the threshold is known, the investment problem is solved, since it is optimal for the firm to wait with the investment when θ is below the threshold θ^* and it is optimal for the firm to make the investment the first time θ is above the threshold. In Appendix A we present a theorem that gives sufficient conditions for the uniqueness of the threshold.

3.   SINGLE SWITCH

In order to find the threshold in the single switch case we first derive an expression for the termination payoff. That is the firm's value at the moment that the firm undertakes its investment. By $V(\zeta)$ we denote the value of the firm when it produces with technology ζ forever, thus

$$V(\zeta) = \int_{t=0}^{\infty} \pi(\zeta) e^{-rt} dt = \frac{\pi(\zeta)}{r}. \qquad (2.9)$$

Equation (2.9) implies that the termination payoff is given by the following expression

$$V(\zeta) - I. \qquad (2.10)$$

The following proposition which is proved in Appendix D gives a sufficient condition for the existence and uniqueness of the threshold θ^*.

PROPOSITION 2.1 *There exists a unique threshold* $\theta^* \in \mathbf{R}_+$ *if the function* π *is concave in* θ.

Given ζ_0 and $\theta(t) = \theta$ the value of the firm is denoted by $F(\theta, \zeta_0)$. Let θ^* be (again) the unique threshold value. From equation (2.10) we derive that in the stopping region, $\{\theta|\theta \geq \theta^*\}$, F is given by

$$F(\theta,\zeta_0) = V(\theta) - I. \qquad (2.11)$$

In the continuation region, $\{\theta|\theta < \theta^*\}$, F must satisfy the following Bellman equation (see Appendix A)

$$rF(\theta,\zeta_0) = \pi(\zeta_0) + \lim_{dt\downarrow 0} \frac{1}{dt} E[dF(\theta,\zeta_0)]. \qquad (2.12)$$

To proceed we must first specify the properties of u, the size of the jump. In Subsection 3.1 we assume it to be constant and in Subsection 3.2 to be stochastic.

3.1   CONSTANT JUMP SIZE

Let us assume that the jump size is constant and equal to some u, $u \geq 0$. Thus $u_i = u$ for $i \in \mathbb{N}$. Given that u is a constant we can split up the continuation region into two parts. In the first part investing is not optimal even after the next jump, i.e. $\{\theta|\theta < \theta^* - u\}$, and in the second part investing is optimal after the next jump: $\{\theta|\theta^* - u \leq \theta < \theta^*\}$.

In the first part of the continuation region, applying Itô's lemma (see Appendix A) gives

$$E[dF(\theta,\zeta_0)] = \lambda dt\,(F(\theta + u,\zeta_0) - F(\theta,\zeta_0)) + o(dt). \qquad (2.13)$$

The definition of $o(dt)$ is stated in Appendix C. Substitution of (2.13) in (2.12) gives

$$rF(\theta,\zeta_0) = \pi(\zeta_0) + \lambda\,(F(\theta + u,\zeta_0) - F(\theta,\zeta_0)). \qquad (2.14)$$

From Proposition 2.4 in Appendix B we know that the solution of equation (2.14) is given by

$$F(\theta,\zeta_0) = c\left(\frac{\lambda}{r+\lambda}\right)^{-\frac{\theta}{u}} + \frac{\pi(\zeta_0)}{r}. \qquad (2.15)$$

To determine the constant c we must use the continuity condition at $\theta = \theta^* - u$. Therefore we need an expression for F in the second part of the continuation region.

The equivalent of equation (2.13) in the second part of the continuation region is

$$E[dF(\theta,\zeta_0)] = \lambda dt\,(V(\theta + u) - I - F(\theta,\zeta_0)) + o(dt). \qquad (2.16)$$

Together with (2.12) this last expression leads to

$$rF(\theta,\zeta_0) = \pi(\zeta_0) + \lambda\,(V(\theta + u) - I - F(\theta,\zeta_0)). \qquad (2.17)$$

Rewriting gives

$$F(\theta, \zeta_0) = \frac{\pi(\zeta_0)}{r+\lambda} + \frac{\lambda}{r+\lambda} (V(\theta+u) - I). \qquad (2.18)$$

Solving the continuity condition at $\theta = \theta^* - u$ gives

$$c = \left(\frac{\lambda}{r+\lambda}\right)^{\frac{\theta^*}{u}} (V(\theta^*) - V(\zeta_0) - I). \qquad (2.19)$$

Summarizing, the value of the firm is given by

$$F(\theta, \zeta_0) = \begin{cases} \left(\frac{\lambda}{r+\lambda}\right)^{\frac{\theta^*-\theta}{u}} (V(\theta^*) - V(\zeta_0) - I) \\ +V(\zeta_0) & \text{if } \theta < \theta^* - u, \\ \frac{\pi(\zeta_0)}{r+\lambda} + \frac{\lambda}{r+\lambda} (V(\theta+u) - I) & \text{if } \theta^* - u \le \theta < \theta^*, \\ V(\theta) - I & \text{if } \theta \ge \theta^*. \end{cases}$$
$$(2.20)$$

In case $\theta < \theta^* - u$ the value of the firm consists of two terms. The second term resembles the value of the firm when the firm produces with technology ζ_0 forever. The first term is the value of the opportunity to invest, in other words the option value. This option value is a multiplication of two parts. The last part is the net present value when the firm adopts technology θ^*. The firm exchanges the profit flow $\pi(\zeta_0)$ for $\pi(\theta^*)$ and pays I for that exchange. The first part of the option value is the discount factor. The investment takes place in the future and therefore the net present value has to be properly discounted. The factor $\frac{\lambda}{r+\lambda}$ is the discounted value of one unit of money that the firms receives after the next technology arrival (see Lemma 2.3 in Appendix D). This factor is raised to the power $\frac{\theta^*-\theta}{u}$, because it takes (continuously spoken) exactly that many arrivals before the firm invests.

The firm is going to switch technologies after the next technology arrival if in the continuation region it holds that $\theta^* - u \le \theta < \theta^*$. This implies that the value of the firm consists of the discounted profit flows generated from now until that technology arrival (first part) and the discounted value of the termination payoff (second part). Notice that in the first part λ is added to the discount rate, so that this is another example of the general notion Dixit and Pindyck, 1996 present on page 87: "… if a profit flow can stop when a Poisson event with arrival rate λ occurs, then we can calculate the expected present value of the stream as if it never stops, but adding λ to the discount rate."

The critical level θ^* is found by solving the value matching condition (see Dixit and Pindyck, 1996) at $\theta = \theta^*$:

$$\frac{\pi\left(\zeta_0\right)}{r+\lambda} + \frac{\lambda}{r+\lambda}\left(\frac{\pi\left(\theta^* + u\right)}{r} - I\right) = \frac{\pi\left(\theta^*\right)}{r} - I. \qquad (2.21)$$

The value matching condition ensures the continuity of the value function at the threshold, i.e. the point where the firm is indifferent between investing right away (right-hand side of equation (2.21)) and investing after the next technology arrival (left-hand side of equation (2.21)).

Define the adoption time t^* as follows

$$t^* = \inf\left(t \mid \theta\left(t\right) \geq \theta^*\right). \qquad (2.22)$$

Since the size of the jump is constant we can calculate after how many technology arrivals the firm is going to switch technologies

$$n^* = \left\lfloor \frac{\theta^* - \theta_0}{u} \right\rfloor + 1, \qquad (2.23)$$

where $\lfloor x \rfloor$ is equal to the integer part of x. Using equation (2.23) it is not hard to see that

$$\Pr\left(t^* \leq t\right) = \sum_{n=n^*}^{\infty} \Pr\left(N\left(t\right) = n\right). \qquad (2.24)$$

The following proposition gives expressions for the expected value and the variance of t^*. The proof is given in Appendix D.

PROPOSITION 2.2 *The expected value and variance of the adoption time* t^* *are equal to*

$$E\left[t^*\right] = \frac{n^*}{\lambda}, \qquad (2.25)$$

$$Var\left[t^*\right] = \frac{n^*}{\lambda^2}. \qquad (2.26)$$

EXAMPLE 2.1 (CONTINUED) *Let us continue our example. Since the profit function is convex, we can not apply Proposition 2.1. However, in Appendix A we show that a sufficient condition for uniqueness is* $\frac{\lambda}{r+\lambda} \leq$ $\left(\frac{\zeta_0}{\zeta_0 + u}\right)^{b-1}$. *Solving equation (2.21) for the parameter set of Farzin et al., 1998, i.e.* $a = \frac{1}{2}$, $p = 200$, $w = 50$, $r = 0.1$, $\lambda = 1$, $u = 0.1$, $\zeta_0 = \theta_0 = 1$ *and* $I = 1600$, *yields* $\theta^* = 2.703$, *so that* $n^* = 18$. *Note that the condition for uniqueness is satisfied for these parameters. With Proposition 2.2* *we find that the expected value of the adoption time is 18 years with a standard deviation of 4.243 years. The left part of Figure 2.2 compares the value of the firm and the termination payoff. The value of the option to invest is plotted in the right part of Figure 2.2, where also the net present value of the investment is plotted.*

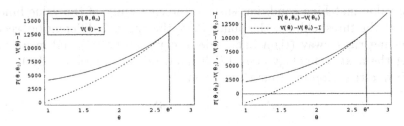

Figure 2.2.   In the left part: value of the firm as function of θ and termination payoff as function of θ. In the right part: value of the option to invest as function of θ and the net present value of the investment as function of θ.

3.2   STOCHASTIC JUMP SIZE

In this subsection we repeat the exercise of Subsection 3.1 for a stochastic jump size. We solve two cases. In the first case we assume that the u_i's are independently and identically distributed according to a uniform distribution and in the second case we assume that the u_i's are independent and identical distributed according to an exponential distribution.

UNIFORM DISTRIBUTION

Let u be distributed according to a uniform distribution on the interval $[0, \overline{u}]$. As in Subsection 3.1 the continuation region is split up into two regions: $\{\theta | \theta < \theta^* - \overline{u}\}$ and $\{\theta | \theta^* - \overline{u} \leq \theta < \theta^*\}$. The Bellman equation for F in the first part of the continuation region is given by

$$rF(\theta, \zeta_0) = \pi(\zeta_0) + \lambda \int_{u=0}^{\overline{u}} (F(\theta + u, \zeta_0) - F(\theta, \zeta_0)) \frac{1}{\overline{u}} du. \qquad (2.27)$$

The solution of (2.27) is (see Proposition 2.5 in Appendix B)

$$F(\theta, \zeta_0) = \gamma_0 e^{\gamma_1 \theta} + \frac{\pi(\zeta_0)}{r}, \qquad (2.28)$$

where γ_0 will be determined by solving the continuity condition at $\theta = \theta^* - \bar{u}$ and γ_1 is the positive solution of the following equation

$$\lambda \left(e^{\bar{u}\gamma_1} - 1 \right) - (r + \lambda)\, \bar{u}\gamma_1 = 0. \tag{2.29}$$

Lemma 2.1 in Appendix B ensures that γ_1 exist and is indeed positive.

The firm is going to switch technologies after the next technology arrival with a positive probability when $\theta(t)$ is in the second part of the continuation region. This gives rise to the following Bellman equation

$$
F(\theta, \varsigma_0) = \frac{\pi(\varsigma_0)}{r + \lambda} + \frac{\lambda}{r + \lambda} \int_{u=0}^{\theta^* - \theta} F(\theta + u, \varsigma_0) \frac{1}{\bar{u}} du
$$

$$
+ \frac{\lambda}{r + \lambda} \int_{u=\theta^* - \theta}^{\bar{u}} \left(V(\theta + u) - I \right) \frac{1}{\bar{u}} du. \tag{2.30}
$$

Proposition 2.6 in Appendix B states that the solution of (2.30) is given by

$$
F(\theta, \varsigma_0) = \frac{\lambda}{(r + \lambda)\bar{u}} \left(h(\theta) - h(\theta^*) \right) e^{-\frac{\lambda\theta}{(r+\lambda)\bar{u}}}
$$

$$
+ \frac{\pi(\varsigma_0)}{r + \lambda} e^{\frac{\lambda(\theta^* - \theta)}{(r+\lambda)\bar{u}}}
$$

$$
+ \frac{\lambda}{(r + \lambda)\bar{u}} e^{\frac{\lambda(\theta^* - \theta)}{(r+\lambda)\bar{u}}} \int_{u=0}^{\bar{u}} \left(V(\theta^* + u) - I \right) du, \tag{2.31}
$$

where $h(\theta)$ is implicitly defined by

$$
\frac{\partial h(\theta)}{\partial \theta} = \left(V(\theta) - I \right) e^{\frac{\lambda\theta}{(r+\lambda)\bar{u}}}. \tag{2.32}
$$

An expression for the constant γ_0 can be found by equating equations (2.28) and (2.31) at $\theta = \theta^* - \bar{u}$.

Summarizing, the value of the firm F is given by

$$
F(\theta, \varsigma_0) =
\begin{cases}
\gamma_0 e^{\gamma_1 \theta} + V(\varsigma_0) & \text{if } \theta < \theta^* - u, \\[2ex]
\frac{\lambda}{(r+\lambda)\bar{u}} \left(h(\theta) - h(\theta^*) \right) e^{-\frac{\lambda\theta}{(r+\lambda)\bar{u}}} \\
+ \frac{\pi(\varsigma_0)}{r+\lambda} e^{\frac{\lambda(\theta^* - \theta)}{(r+\lambda)\bar{u}}} + \frac{\lambda}{(r+\lambda)\bar{u}} e^{\frac{\lambda(\theta^* - \theta)}{(r+\lambda)\bar{u}}} \\
\times \int_{u=0}^{\bar{u}} \left(V(\theta^* + u) - I \right) du & \text{if } \theta^* - u \leq \theta < \theta^*, \\[2ex]
V(\theta) - I & \text{if } \theta \geq \theta^*.
\end{cases}
\tag{2.33}
$$

The critical level θ^* is found by solving the value matching condition at $\theta = \theta^*$:

$$\frac{\pi(\zeta_0)}{r+\lambda} + \frac{\lambda}{(r+\lambda)\bar{u}} \int\limits_{u=0}^{\bar{u}} (V(\theta^* + u) - I)\, du = \frac{\pi(\theta^*)}{r} - I. \qquad (2.34)$$

We can not give an equivalent of Proposition 2.2 in this case, because there does not exist a closed form expression for the n-th fold convolution of the uniform distribution.

EXAMPLE 2.1 (CONTINUED) *In Appendix A we show that the threshold is unique if $\zeta_0 \geq \bar{u}$, which is satisfied for our parameters. Solving equation (2.34) for our parameter set and $\bar{u} = 0.2$, thus $E[u_i] = 0.1$ and $Var[u_i] = 0.00333$, yields $\theta^* = 2.713$. Thus introducing the uncertainty in the size of the jump causes a very small change in the threshold, namely 0.36 percent. Using simulation we calculated that the expected value and the standard deviation of the adoption are equal to 17.79 years and 4.87 years, respectively.*

EXPONENTIAL DISTRIBUTION

The main difference with the analysis above is that in this case for every θ in the continuation region there is a positive probability that the next jump is large enough to lift θ above the threshold. Therefore we do not have to split the continuation region in two parts. The following Bellman equation must hold in the continuation region

$$F(\theta, \zeta_0) = \frac{\pi(\zeta_0)}{r+\lambda} + \frac{\lambda}{r+\lambda} \int\limits_{u=0}^{\theta^*-\theta} F(\theta + u, \zeta_0)\, \mu e^{-\mu u}\, du$$

$$+ \frac{\lambda}{r+\lambda} \int\limits_{u=\theta^*-\theta}^{\infty} (V(\theta + u) - I)\, \mu e^{-\mu u}\, du. \qquad (2.35)$$

Applying Proposition 2.8 (stated in Appendix B) to equation (2.35) gives

$$F(\theta, \zeta_0) = \frac{\lambda}{r+\lambda} e^{\frac{\mu(r\theta + \lambda\theta^*)}{r+\lambda}} \int\limits_{u=\theta^*}^{\infty} (V(u) - I)\, \mu e^{-\mu u}\, du$$

$$+ \left(1 - \frac{\lambda}{r+\lambda} e^{\frac{\mu r(\theta - \theta^*)}{r+\lambda}}\right) V(\zeta_0). \qquad (2.36)$$

Expressions (2.10) and (2.36) together give the value of the firm

$$
F\left(\theta,\zeta_0\right) =
\begin{cases}
\dfrac{\lambda}{r+\lambda}e^{\frac{\mu(r\theta+\lambda\theta^*)}{r+\lambda}}\displaystyle\int\limits_{u=\theta^*}^{\infty}\left(V\left(u\right)-I\right)\mu e^{-\mu u}du \\[2mm]
+\left(1-\dfrac{\lambda}{r+\lambda}e^{\frac{\mu r(\theta-\theta^*)}{r+\lambda}}\right)V\left(\zeta_0\right) & \text{if } \theta < \theta^*, \\[2mm]
V\left(\theta\right)-I & \text{if } \theta \geq \theta^*.
\end{cases}
$$
$$(2.37)$$

The critical level θ^* is found by solving the value matching condition at $\theta = \theta^*$:

$$
\frac{\pi\left(\zeta_0\right)}{r+\lambda}+\frac{\lambda}{r+\lambda}e^{\mu\theta^*}\int\limits_{u=\theta^*}^{\infty}\left(V\left(u\right)-I\right)\mu e^{-\mu u}du = \frac{\pi\left(\theta^*\right)}{r}-I. \qquad (2.38)
$$

PROPOSITION 2.3 *The expected value and the variance of the adoption time t^* are equal to*

$$
E\left[t^*\right] = \frac{\mu\left(\theta^*-\theta_0\right)+1}{\lambda}, \qquad (2.39)
$$

$$
Var\left[t^*\right] = \frac{2\mu\left(\theta^*-\theta_0\right)+1}{\lambda^2}. \qquad (2.40)
$$

EXAMPLE 2.1 (CONTINUED) *Let $\mu = 10$ then $E\left[u_i\right] = Var\left[u_i\right] = 0.1$, $i \in \mathbb{N}$. In Appendix A we derived that the threshold is unique if $\zeta_0^{b-1} \geq \frac{\lambda}{r}\frac{\Gamma(b)}{\mu^{b-1}}$, which is satisfied for our parameters. Solving equation (2.38) gives $\theta^* = 2.732$. Table 2.1 shows the thresholds and the expected value and standard deviation of the adoption times of the three cases. We conclude that the distribution of the size of the jump does not matter very much. Huisman, 1996 showed that there are also hardly any changes when the interarrival times are constant instead of exponentially distributed. The standard deviation of the adoption time in the exponential jump size case is the largest because the variance of the jump size is the largest in that case.*

4.   MULTIPLE SWITCHES

In this section we solve the technology investment problem if the firm can switch technologies n times. One of the conclusions of the last section is that the probability distribution of the size of the jump, degenerate, uniform or exponential, does not influence the outcome very much. Therefore, in this section, we only explicitly solve the model for the case that the jump size is constant. After that we discuss the stochastic case.

Table 2.1. Thresholds, expected value and standard deviation of the adoption time in the three different cases.

u_i's	θ^*	$E[t^*]$	$Sd[t^*]$
constant	2.703	18.00	4.24
uniform	2.713	17.79	4.87
exponential	2.732	18.32	5.97

4.1   CONSTANT JUMP SIZE

The investment problem of the firm consists of n optimal stopping problems, where the outcome of the i-th optimal stopping problem, the threshold θ_i^*, is an input for the $(i+1)$-th optimal stopping problem. The n-th optimal stopping problem has already been solved in the previous section.

The value of the firm before the last (the n-th) technology switch is denoted by $F_n(\theta, \zeta_{n-1})$ and equals (cf. equation (2.20)),

$$
F_n(\theta, \zeta_{n-1}) = \begin{cases} \left(\frac{\lambda}{r+\lambda}\right)^{\frac{\theta_n^*-\theta}{u}} (V(\theta_n^*) - I) \\ \quad + \left(1 - \left(\frac{\lambda}{r+\lambda}\right)^{\frac{\theta_n^*-\theta}{u}}\right) V(\zeta_{n-1}) & \text{if } \theta < \theta_n^* - u, \\ \frac{\pi(\zeta_{n-1})}{r+\lambda} + \frac{\lambda}{r+\lambda}(V(\theta+u) - I) & \text{if } \theta_n^* - u \leq \theta < \theta_n^*, \\ V(\theta) - I & \text{if } \theta \geq \theta_n^*. \end{cases}
$$

$$(2.41)$$

From the analysis of the previous section we know that the threshold θ_n^* is defined as the solution of the following equation

$$
\frac{\pi(\zeta_{n-1})}{r+\lambda} + \frac{\lambda}{r+\lambda}(V(\theta_n^* + u) - I) = V(\theta_n^*) - I. \qquad (2.42)
$$

Next let us analyze the i-th ($i \in \{1, \ldots, n-1\}$) optimal stopping problem. The value of the firm when it is about to make its i-th technology switch is given by

$$
F_i\left(\theta, \zeta_{i-1}\right) =
\begin{cases}
\left(\frac{\lambda}{r+\lambda}\right)^{\frac{\theta_i^* - \theta}{u}} \left(F_{i-1}\left(\theta_i^*, \theta_i^*\right) - I\right) \\
+ \left(1 - \left(\frac{\lambda}{r+\lambda}\right)^{\frac{\theta_i^* - \theta}{u}}\right) V\left(\zeta_{i-1}\right) & \text{if } \theta < \theta_i^* - u, \\
\frac{\pi(\zeta_{i-1})}{r+\lambda} \\
+ \frac{\lambda}{r+\lambda}\left(F_{i-1}\left(\theta + u, \theta + u\right) - I\right) & \text{if } \theta_i^* - u \le \theta < \theta_i^*, \\
F_{i-1}\left(\theta, \theta\right) - I & \text{if } \theta \ge \theta_i^*.
\end{cases}
$$
$$(2.43)$$

The threshold θ_i^* is the solution of the value matching condition for F_i at $\theta = \theta_i^*$:

$$
\frac{\pi\left(\zeta_{i-1}\right)}{r+\lambda} + \frac{\lambda}{r+\lambda}\left(F_{i-1}\left(\theta_i^* + u, \theta_i^* + u\right) - I\right) = F_{i-1}\left(\theta_i^*, \theta_i^*\right) - I. \quad (2.44)
$$

The following theorem states how the n thresholds are calculated.

THEOREM 2.1 *The thresholds θ_i^*, $i \in \{1, \dots, n\}$ are found by simultaneously solving the following set of equations*

$$
\begin{aligned}
\frac{\pi\left(\zeta_{i-1}\right)}{r+\lambda} + \frac{\lambda}{r+\lambda}\left(F_{i-1}\left(\theta_i^* + u, \theta_i^* + u\right) - I\right) &= F_{i-1}\left(\theta_i^*, \theta_i^*\right) - I, \\
\frac{\pi\left(\zeta_{n-1}\right)}{r+\lambda} + \frac{\lambda}{r+\lambda}\left(V\left(\theta_n^* + u\right) - I\right) &= V\left(\theta_n^*\right) - I,
\end{aligned}
$$

where $\zeta_i = \inf\left(\theta_j | \theta_j \ge \theta_i^* \text{ and } j \in \mathbb{N}_0\right)$, $i \in \{1, \dots, n-1\}$, and the functions F_i and V are defined by equations (2.43) and (2.9), respectively.

Although the theorem tells us what equations we should solve in order to find the solution to the technology investment problem, it seems impossible to do so in practice. The problem is caused by the ζ_i's. When the jump size is constant it holds that $\zeta_i = \left\lfloor \frac{\theta_i^*}{u} \right\rfloor + 1$, $i \in \{1, \dots, n-1\}$, but it is impossible to solve the system of equations after substitution of these expressions.

In case the jump size is constant the technology investment problem can be solved in the following way. Therefore we introduce some more notation. Let j denote the number of the technology currently in use by the firm and k the number of the best technology available, i.e. at time t we have $\zeta(t) = \theta_j$ and $\theta(t) = \theta_k$. Define the following functions

$$
g_i(j, k) =
\begin{cases}
\max\limits_{m \ge k} f_{i+1}(j, m, k) & \text{if } i \in \{0, \dots, n-1\}, \\
V(\theta_j) & \text{if } i = n,
\end{cases}
\quad (2.45)
$$

and for $i \in \{0, \dots, n-1\}$,

$$f_{i+1}(j, m, k) = \left(1 - \left(\frac{\lambda}{r+\lambda}\right)^{m-k}\right) V(\theta_j)$$
$$+ \left(\frac{\lambda}{r+\lambda}\right)^{m-k} g_{i+1}(m, m). \qquad (2.46)$$

The function g gives the value of the firm as function of i, j, and k, where i equals the number of switches the firm has made so far. When the firm's i-th technology switch is from technology θ_j to technology θ_m and the current best technology is θ_k, the value of the firm is given by $f_i(j, m, k)$. Note that

$$f_{i+1}(j, m, k) = V(\theta_j) + E\left[e^{-r(T_m - t)} \middle| t \in [T_k, T_{k+1})\right]$$
$$\times (g_{i+1}(m, m) - V(\theta_j)). \qquad (2.47)$$

From Lemma 2.3 (see Appendix D) we know that

$$E\left[e^{-r(T_m - t)} \middle| t \in [T_k, T_{k+1})\right] = \left(\frac{\lambda}{r+\lambda}\right)^{m-k}. \qquad (2.48)$$

Substituting this last equation into equation (2.47) gives equation (2.46). Taking all this into account brings us to the following theorem.

THEOREM 2.2 *Let the efficiencies levels θ_i, $i \in \mathbb{N}_0$ be given and set $\zeta_0 = \theta_0$. Then it is optimal for the firm to adopt the technologies with the following efficiency levels at the moment that these technologies become available:*

$$\zeta_i = \theta_{m_i^*}, \quad i \in \{1, \dots, n\},$$

in which

$$m_i^* = \arg \max_{m \geq m_{i-1}^*} \left(f_{i-1}(m_{i-1}^*, m, m_{i-1}^*)\right), \quad i \in \{1, \dots, n\},$$

where $m_0^* = 0$. The value of the firm equals $g_0(0, 0)$.

Note that Theorem 2.2 can be applied for any given set of efficiency levels, i.e. the jump sizes do not have to be constant.

EXAMPLE 2.1 (CONTINUED) *Table 2.2 gives the results of applying Theorem 2.2 to our parameter set and $n \in \{1, 2, 3, 4, 5\}$. From that table we conclude that increasing the firm's flexibility, i.e. the ability of making more technology switches, increases the firm's value. Further we see that the more switches the firm can make, the earlier the first switch is made.*

Table 2.2. Value of the firm and efficiencies of technologies adopted for $n \in \{1, 2, 3, 4, 5\}$.

n	1	2	3	4	5
$g_n(0,0)$	4172.69	5181.28	5722.92	6041.76	6236.30
ζ_1	2.8	2.3	2.2	2.1	2.0
ζ_2	-	3.7	3.2	3.0	2.8
ζ_3	-	-	4.5	4.0	3.6
ζ_4	-	-	-	5.3	4.5
ζ_5	-	-	-	-	5.8

4.2   STOCHASTIC JUMP SIZE

Theorem 2.1 also holds in case the jump size is stochastic. The only things that change are the functions F_i. In the stochastic jump case the values of ζ_i, $i \in \{1, \dots, n-1\}$, are not known beforehand. However, to obtain a solution the efficiency parameters of the technologies that arrive in the future must be known. Therefore, the best that can be done is to design an approximation of the solution. For instance, the following approach can be chosen. In order to obtain approximations of the thresholds set $\zeta_i = \theta_i^*$, $i \in \{1, \dots, n-1\}$. Of course, after each technology adoption we know the exact value of that ζ_i and we update the new approximations.

Hence, when we replace $V(\theta_j)$ by $E[V(\theta_j(t))|t]$ in equations (2.45) and (2.46), Theorem 2.2 can be used to make a prediction about the efficiencies of the technologies that the firm should adopt in case the jump size is stochastic. However after each technology arrival these predictions must be corrected.

Now we are in a position to point out a mistake in Farzin et al., 1998 (see also Doraszelski, 2001). The corresponding equation to equation (2.44) in case the jump size is uniformly distributed is given by

$$\frac{\pi(\zeta_{i-1})}{r+\lambda} + \frac{\lambda}{(r+\lambda)\overline{u}} \int\limits_{u=0}^{\overline{u}} (F_{i-1}(\theta_i^* + u, \theta_i^* + u) - I)\, du$$

$$= F_{i-1}(\theta_i^*, \theta_i^*) - I. \tag{2.49}$$

Neglecting the possibility that the firm is going to adopt successive technologies, i.e. $\theta_i^* + \bar{u} < \theta_{i+1}^*$, it holds that (cf. equation (2.27))

$$F_{i-1}\left(\theta_i^*, \theta_i^*\right) = \frac{\pi\left(\theta_i^*\right)}{r+\lambda} + \frac{\lambda}{(r+\lambda)\,\bar{u}} \int_{u=0}^{\bar{u}} F_{i-1}\left(\theta_i^* + u, \theta_i^*\right) du. \qquad (2.50)$$

Substitution of (2.50) into (2.49) gives

$$\frac{\pi\left(\zeta_{i-1}\right)}{r} + \frac{\lambda}{r\bar{u}} \int_{u=0}^{\bar{u}} \left(F_{i-1}\left(\theta_i^* + u, \theta_i^* + u\right) - F_{i-1}\left(\theta_i^* + u, \theta_i^*\right)\right) du$$

$$= \frac{\pi\left(\theta_i^*\right)}{r} - I. \qquad (2.51)$$

Due to the fact that Farzin et al., 1998 did not take into account the dependence of F on ζ, i.e. the efficiency of the technology that the firm currently uses, they cancelled out the integral on the left-hand side of equation (2.51). Since the value of the firm is increasing in the efficiency of the technology in use, we conclude that the optimal triggers are larger than the ones derived in Section 4 of Farzin et al., 1998. Notice that the integral on the left-hand side of equation (2.51) resembles the value of the option to invest.

5.   NET PRESENT VALUE METHOD

The net present value method states that an investment should be made when the present value of the cash flows generated by that investment exceeds the investment cost. The net present value method implicitly assumes that an investment is either reversible or if irreversible it is a now or never opportunity. Dixit and Pindyck, 1996 extensively discuss that most investment problems do not satisfy these assumptions. In the technology adoption framework it is clear that the investment is irreversible. As an example think of buying a personal computer. Clearly it is possible to postpone such an investment. Therefore for applying the net present value the investment should be reversible. In the case of a personal computer this is (almost surely) not true, i.e. the investment is irreversible.

To incorporate the irreversibility and the possibility to delay an investment, the real option theory was developed. For a good introduction and overview we refer to Dixit and Pindyck, 1996. In the real option theory investment opportunities are looked upon as options. The firm has the right but not the obligation to make the investment. At the moment that the firm invests the option is killed and because the option is valuable the firm looses money. Therefore the lost option value should be incorporated in the investment analysis. In the technology investment problem the option to invest is valuable, because with positive probability the firm can buy a better technology for the same amount of money if it waits just a little with making the investment.

The following theorem states what technologies the firm should adopt if the firm uses the net present value method to solve the investment problem.

THEOREM 2.3 *According to the net present value method the firm adopts the technologies with the following efficiencies:*

$$\zeta_i^{NPV} = \inf\left(\theta_j \mid \theta_j \geq \theta_i^{NPV} \text{ and } j \in \mathbb{N}_0\right), \quad i \in \{1, \dots, n\},$$

where θ_i^{NPV}, $i \in \{1, \dots, n\}$, *is the solution of*

$$V\left(\theta_i^{NPV}\right) - I = V\left(\zeta_{i-1}^{NPV}\right), \tag{2.52}$$

and $\zeta_0^{NPV} = \zeta_0$.

The theorem is easily verified by looking at equation (2.9). Comparing equations (2.52) and (2.51) yields the following corollary.

COROLLARY 2.1 *The net present value prescribes the firm to make the investments too early, i.e.* $\zeta_i^{NPV} < \zeta_i$, $i \in \{1, \dots, n\}$.

EXAMPLE 2.1 (CONTINUED) *We calculated the efficiencies of the first five technologies that should be adopted according to the net present value method. The results are put in Table 2.3. Comparing this table with the last column of Table 2.2 we see that indeed* $\zeta_i^{NPV} < \zeta_i$ *for* $i \in \{1, 2, 3, 4, 5\}$.

Table 2.3.   Efficiencies of the first five technologies that should be adopted according to the net present value method.

i	ζ_i^{NPV}
1	1.4
2	1.7
3	2.0
4	2.2
5	2.4

6.   CONCLUSIONS

In this chapter we analyzed the technology investment problem of a single firm. New technologies arrive according to a Poisson jump process and the efficiency increases were modelled in three ways. It turns out that the probability distribution of the jump size (degenerate, uniform, or exponential) does not influence the result very much (see Example 2.1). This conclusion holds generally whenever the expected value of investing after the next technology arrival is almost the same for each approach.

It is only possible to completely solve the multiple switch case at the beginning of the planning period if the efficiencies of the new technologies are known beforehand. This implies that in the stochastic jump size case the solution of the model must be updated after each technology arrival.

In Section 5 the incorrectness of the net present value method is proved. The example showed that there is a significant difference between the optimal adoption pattern and the one proposed by the net present value method.

In the next chapter the model is extended by making the investment costs decreasing over time.

Appendices
A.   OPTIMAL STOPPING

Consider the following dynamic problem of a risk-neutral and value maximizing firm that discounts against rate $r\,(>0)$. The firm has the opportunity to undertake a project. The value of the project depends on one state variable, $x\,(t) \in \mathbf{R}$, which evolves stochastically over time $t\,(\geq 0)$ according to an Itô process or a Poisson jump process.

DEFINITION 2.1 *The stochastic variable $x\,(t)$ behaves according to an Itô process if and only if for $t \geq 0$*

$$dx\,(t) = f\,(t, x\,(t))\,dt + g\,(t, x\,(t))\,d\omega\,(t)\,,$$
$$x\,(0) = x_0,$$

where $d\omega\,(t)$ is an increment of a Wiener process, i.e. $d\omega\,(t)$ is distributed according to a normal distribution with mean 0 and variance dt, and $x_0 \in \mathbf{R}$.

DEFINITION 2.2 *The stochastic variable $x\,(t)$ behaves according to a Poisson jump process if and only if for $t \geq 0$*

$$dx\,(t) = \begin{cases} u\,(t, x\,(t)) & \text{with probability } \lambda dt, \\ 0 & \text{with probability } 1 - \lambda dt, \end{cases}$$
$$x\,(0) = x_0,$$

where $u(t, x(t))$ *is deterministic or distributed according to some probability distribution and* $x_0 \in \mathbf{R}$.

Let $\Omega(x(t))$, with $\Omega : \mathbf{R} \to \mathbf{R}$, be the termination payoff when the firm undertakes the project at state $x(t)$. Before the project is undertaken the firm receives a profit flow $\pi(x(t))$, with $\pi : \mathbf{R} \to \mathbf{R}$. This profit flow stops at the moment that the firm undertakes the project. This investment problem is called an optimal stopping problem (see also Chapter 4 of Dixit and Pindyck, 1996).

Denote the value of the project before stopping by $F(x(t))$. Then given state $x(t)$ and given that the firm has not stopped before, the value of the project is given by

$$F(x(t)) = \max\left(\Omega(x(t)), \pi(x(t))\,dt + e^{-rdt}E\left[F(x(t+dt))|\,x(t)\right]\right).$$
$$(2.53)$$

The first argument within the maximization operator is equal to the value of stopping at time $t\,(\geq 0)$. The second argument equals the value of not stopping at time t, and acting optimally from time $t+dt$ onwards. This is called the Bellman principle of optimality. In the continuation region the second argument within the maximization operator is the largest. This implies that F must satisfy the following so-called Bellman equation in the continuation region

$$rF(x(t)) = \pi(x(t)) + \lim_{dt\downarrow 0}\frac{1}{dt}E\left[dF(x(t))\right].$$
$$(2.54)$$

The expectation in equation (2.54) can be calculated with Itô's lemma.

Itô's lemma *Let* $x(t)$ *behave according to an Itô process or a Poisson jump process and let* $G(t, x(t))$ *be a function that is once differentiable with respect to* t *and twice differentiable with respect to* $x(t)$. *Then*

$$dG(t, x(t)) = \frac{\partial G(t, x(t))}{\partial t}dt + \frac{\partial G(t, x(t))}{\partial x(t)}dx(t)$$
$$+ \frac{1}{2}\frac{\partial^2 G(t, x(t))}{\partial (x(t))^2}(dx(t))^2 + o(dt).$$

For deterministic processes as well as for Poisson process the term $(dx(t))^2$ is of the order $(dt)^2$ and can therefore be added to the $o(dt)$ term. Since $(dx(t))^2$ is of order dt for Itô processes it is stated explicitly in Itô's lemma. For a proof of Itô's lemma we refer the interested reader to Karatzas and Shreve, 1991.

Suppose that the termination payoff is increasing in $x(t)$ and that the profit flow is constant. Then intuition suggests that there exists a threshold x^* such that undertaking the project is optimal when $x(t) > x^*$ and waiting is optimal if $x(t) < x^*$. Hence, the stopping region is defined to be equal to $\{x \in \mathbf{R} \,|\, x \geq x^*\}$ and the continuation region is given by $\{x \in \mathbf{R} \,|\, x < x^*\}$.

Let $F(x(t))$ be the solution of equation (2.54). Then the threshold x^* is found by solving the value matching condition at x^* :

$$F(x^*) = \Omega(x^*). \tag{2.55}$$

Whenever the $x(t)$ process can pass the threshold continuously the smooth pasting condition must hold at x^* :

$$\left.\frac{\partial F(x(t))}{\partial x(t)}\right|_{x(t)=x^*} = \left.\frac{\partial \Omega(x(t))}{\partial x(t)}\right|_{x(t)=x^*}. \tag{2.56}$$

The interested reader is referred to Dixit, 1991 and Dixit, 1993 for a more rigorous treatment of smooth pasting and the control of Brownian motion.

The following theorem gives sufficient conditions for the uniqueness of this threshold. The proof follows Appendix 4.B of Dixit and Pindyck, 1996.

THEOREM 2.4 *Let* $\Phi(x(t+dt)|x(t))$, *with* $\Phi : D \to [0,1]$ *and* $D \subseteq \mathbf{R}$, *be a cumulative probability distribution function such that*

$$E[g(x(t+dt))|x(t)] = \int_{y \in D} g(y)\, d\Phi(y|x(t)),$$

where $g(y)$, *with* $g : D \to \mathbf{R}$, *is a given function. Then given that there exists a threshold* x^*, *this threshold is unique if the following two conditions are satisfied.*

1. *The function* $\pi(x(t)) - r\Omega(x(t)) + \lim_{dt \downarrow 0} \frac{1}{dt} E[d\Omega(x(t))|x(t)]$ *is decreasing in* $x(t)$.

2. *There is positive persistence of uncertainty, i.e. let* $x_1 < x_2$ *then it holds for all* $y \in D$ *that* $\Phi(y|x_1) > \Phi(y|x_2)$.

PROOF OF THEOREM 2.4 *Define*

$$G(x(t)) = F(x(t)) - \Omega(x(t)), \tag{2.57}$$

then

$$G\left(x\left(t\right)\right)$$

$$= \max\left(0, \pi\left(x\left(t\right)\right)dt - \Omega\left(x\left(t\right)\right) + e^{-rdt}E\left[F\left(x\left(t+dt\right)\right)\right|\left.x\left(t\right)\right]\right)$$

$$= \max(0, \pi\left(x\left(t\right)\right)dt - \Omega\left(x\left(t\right)\right) + e^{-rdt}E\left[\Omega\left(x\left(t+dt\right)\right)\right|\left.x\left(t\right)\right]$$
$$+e^{-rdt}E\left[G\left(x\left(t+dt\right)\right)\right|\left.x\left(t\right)\right]). \qquad (2.58)$$

Given the two conditions the function G must be decreasing in $x\left(t\right)$. Consider the second part within the maximization operator. The first condition ensures that the first three arguments together are decreasing in $x\left(t\right)$. If G is decreasing in $x\left(t\right)$, the fourth argument is also decreasing in $x\left(t\right)$, because of the second condition. This implies that, given a decreasing function G, the right-hand side of (2.58) is again a decreasing function. A higher $x\left(t\right)$ shifts the probability distribution Φ uniformly to the right and therefore the expected value decreases. Thus G is decreasing in $x\left(t\right)$ which implies that, given that there exists a threshold x^*, this threshold is unique. □

Note that Theorem 2.4 does not guarantee the existence of a threshold, but if we can derive a threshold and the conditions are satisfied this threshold is unique. On the other hand, Theorem 2.4 provides sufficiency conditions which are by no means necessary. Next we rewrite the conditions in Theorem 2.4 for two specific cases. In the first case $x\left(t\right)$ follows a Brownian motion and in the second case $x\left(t\right)$ follows a Poisson jump process.

A.1   BROWNIAN MOTION PROCESS

Let $x\left(t\right)$ follow a Brownian motion with parameters μ and σ for $t \geq 0$, i.e.

$$dx\left(t\right) = \mu dt + \sigma d\omega\left(t\right), \qquad (2.59)$$
$$x\left(0\right) = x_0, \qquad (2.60)$$

with $d\omega\left(t\right)$ the increment of a Wiener process. Thus $d\omega\left(t\right)$ is distributed according to a normal distribution with mean 0 and variance dt, which implies that $dx\left(t\right)$ is distributed according to a normal distribution with mean μdt and variance $\sigma^2 dt$. In this case the cumulative probability distribution Φ is equal to

$$\Phi\left(y\right|\left.x\right) = \int_{z=-\infty}^{y} \frac{1}{\sigma\sqrt{dt}\sqrt{2\pi}} e^{-\frac{\left(z-\mu dt-x\right)^2}{\sigma^2 dt}} dz. \qquad (2.61)$$

From the last equation it follows that the second condition of Theorem 2.4 is satisfied if $x(t)$ follows a Brownian motion. Now, let us turn to the first condition. Expanding $E\left[d\Omega\left(x\left(t\right)\right)|x\left(t\right)\right]$ with Itô's lemma gives

$$E\left[d\Omega\left(x\left(t\right)\right)|x\left(t\right)\right] = \mu\frac{\partial\Omega\left(x\left(t\right)\right)}{\partial x\left(t\right)}dt + \frac{1}{2}\sigma^2\frac{\partial^2\Omega\left(x\left(t\right)\right)}{\partial\left(x\left(t\right)\right)^2}dt + o\left(dt\right).$$

(2.62)

Uniqueness of the threshold is guaranteed if the function

$$\pi\left(x\left(t\right)\right) - r\Omega\left(x\left(t\right)\right) + \mu\frac{\partial\Omega\left(x\left(t\right)\right)}{\partial x\left(t\right)} + \frac{1}{2}\sigma^2\frac{\partial^2\Omega\left(x\left(t\right)\right)}{\partial\left(x\left(t\right)\right)^2}$$

(2.63)

is decreasing in $x\left(t\right)$.

A.2   POISSON JUMP PROCESS

Here we assume that $x\left(t\right)$ behaves according to the following Poisson jump process for $t \geq 0$:

$$dx\left(t\right) = \begin{cases} u & \text{with probability } \lambda dt, \\ 0 & \text{with probability } 1 - \lambda dt, \end{cases}$$

(2.64)

$$x\left(0\right) = x_0,$$

(2.65)

where u is distributed according to some probability distribution with density function ϕ, with $\phi : S \rightarrow [0,1]$ and $S \subseteq \mathbf{R}_+$. Let $\underline{z} = \inf_{w \in S}\left(w\right)$ and $\overline{z} = \sup_{w \in S}\left(w\right)$. The cumulative probability distribution Φ is given by

$$\Phi\left(y|x\right) = \begin{cases} 0 & \text{if } y - x < \underline{z}, \\ \int_{z=\underline{z}}^{y-x}\phi\left(z\right)dz & \text{if } \underline{z} \leq y - x < \overline{z}, \\ 1 & \text{if } y - x \geq \overline{z}. \end{cases}$$

(2.66)

Hence, the second condition of Theorem 2.4 is satisfied in this case. Applying Itô's lemma gives for this case

$$E\left[d\Omega\left(x\left(t\right)\right)|x\left(t\right)\right] = \lambda dt\int_{z \in S}\left(\Omega\left(x\left(t\right) + z\right) - \Omega\left(x\left(t\right)\right)\right)\phi\left(z\right)dz + o\left(dt\right).$$

(2.67)

This implies that the first condition of Theorem 2.4 is that the following function should be decreasing in $x\left(t\right)$

$$\pi\left(x\left(t\right)\right) - \left(r + \lambda\right)\Omega\left(x\left(t\right)\right) + \lambda\int_{z \in S}\Omega\left(x\left(t\right) + z\right)\phi\left(z\right)dz.$$

(2.68)

Now we can derive sufficient conditions for the uniqueness of the threshold in Example 2.1. In that example we have

$$\pi(\theta) = \varphi\zeta_0^b, \tag{2.69}$$

$$\Omega(\theta) = \frac{\varphi\theta^b}{r} - I. \tag{2.70}$$

First let the jumps be constant, then equation (2.68) becomes

$$\varphi\zeta_0^b - (r+\lambda)\left(\frac{\varphi\theta^b}{r} - I\right) + \lambda\left(\frac{\varphi(\theta+u)^b}{r} - I\right). \tag{2.71}$$

This last equation is decreasing in θ if and only if

$$-(r+\lambda)b\theta^{b-1} + \lambda b(\theta+u)^{b-1} < 0. \tag{2.72}$$

Rewriting equation (2.72) gives

$$\frac{\lambda}{r+\lambda} < \left(\frac{\theta}{\theta+u}\right)^{b-1}, \tag{2.73}$$

which is strongest for $\theta = \theta_0$. Therefore, given the fact that $\theta_0 = \zeta_0$, equation (2.72) holds if and only if

$$\frac{\lambda}{r+\lambda} \leq \left(\frac{\zeta_0}{\zeta_0+u}\right)^{b-1}. \tag{2.74}$$

When the jumps are uniformly distributed a sufficient condition for uniqueness is $\zeta_0 \geq \bar{u}$. To see this, note that in this case equation (2.68) equals

$$\varphi\zeta_0^b - (r+\lambda)\left(\frac{\varphi\theta^b}{r} - I\right) + \frac{\lambda}{\bar{u}(b+1)}\left(\frac{\varphi(\theta+u)^{b+1}}{r} - \frac{\varphi\theta^{b+1}}{r} - I\right), \tag{2.75}$$

which is decreasing in θ for $\zeta_0 \geq \bar{u}$ since

$$-(r+\lambda)b\theta^{b-1} + \frac{\lambda}{\bar{u}}\left((\theta+\bar{u})^b - \theta^b\right)$$
$$\leq -(r+\lambda)\theta^{b-1} + \frac{\lambda}{\bar{u}}\left(\theta^b + \bar{u}^b - \theta^b\right)$$
$$= -(r+\lambda)\theta^{b-1} + \lambda\bar{u}^{b-1}$$
$$< -\lambda\zeta_0^{b-1} + \lambda\bar{u}^{b-1}$$
$$\leq 0.$$

Lastly we derive a sufficient condition for the case that the jumps are exponentially distributed. The equivalent of equation (2.68) is

$$\varphi \zeta_0^b - (r+\lambda)\left(\frac{\varphi \theta^b}{r} - I\right) + \lambda \int_{z=0}^{\infty}\left(\frac{\varphi(\theta+z)^b}{r} - I\right)\mu e^{-\mu z}dz. \quad (2.76)$$

Differentiating equation (2.76) with respect to θ and multiplying by $\frac{r}{\varphi b}$ gives

$$-(r+\lambda)\theta^{b-1} + \lambda \int_{z=0}^{\infty}(\theta+z)^{b-1}\mu e^{-\mu z}dz$$

$$\leq -(r+\lambda)\theta^{b-1} + \lambda \int_{z=0}^{\infty}\theta^{b-1}\mu e^{-\mu z}dz + \lambda \int_{z=0}^{\infty}z^{b-1}\mu e^{-\mu z}dz$$

$$= -r\theta^{b-1} + \lambda\frac{\Gamma(b)}{\mu^{b-1}}\int_{z=0}^{\infty}\frac{\mu^b}{\Gamma(b)}z^{b-1}e^{-\mu z}dz$$

$$= -r\theta^{b-1} + \lambda\frac{\Gamma(b)}{\mu^{b-1}},$$

where $\Gamma(b)$ is defined by equation (2.111) in Appendix C. Thus the threshold is unique if

$$\zeta_0^{b-1} \geq \frac{\lambda}{r}\frac{\Gamma(b)}{\mu^{b-1}}. \quad (2.77)$$

B.   DIFFERENTIAL EQUATIONS

PROPOSITION 2.4 *Let the constants a_0, a_1, and a_2 be positive constants. The solution of*

$$f(x) = a_0 + a_1 f(x+a_2), \quad (2.78)$$

is given by

$$f(x) = c(a_1)^{-\frac{x}{a_2}} + \frac{a_0}{1-a_1}, \quad (2.79)$$

where c is a constant to be determined by some boundary condition.

PROOF OF PROPOSITION 2.4 *The correctness of the proposition is easily verified after substitution of equation (2.79) into equation (2.78).* □

PROPOSITION 2.5 *Let the constants a_0, a_1, and a_2 be positive constants. The solution of*

$$f(x) = a_0 + a_1 \int_{y=0}^{a_2} f(x+y)\, dy, \qquad (2.80)$$

is given by

$$f(x) = c_0 e^{c_1 x} + \frac{a_0}{1 - a_1 a_2}, \qquad (2.81)$$

where c_1 is the solution of

$$a_1 \left(e^{a_2 c_1} - 1\right) - c_1 = 0, \qquad (2.82)$$

and c_0 is a constant to be determined by some boundary condition.

PROOF OF PROPOSITION 2.5 *Substitution of equation (2.81) in equation (2.80) and rearranging gives equation (2.82).* $\square$

LEMMA 2.1 *Let a_1 and a_2 be positive constants. Equation (2.82) has a unique positive solution c_1 if and only if $a_1 a_2 < 1$.*

PROOF OF LEMMA 2.1 *Define*

$$g(c_1) = a_1 \left(e^{a_2 c_1} - 1\right) - c_1 = 0.$$

Then the first and second derivative of g are given by

$$\frac{\partial g(c_1)}{\partial c_1} = a_1 a_2 e^{a_2 c_1} - 1,$$

and

$$\frac{\partial^2 g(c_1)}{\partial c_1^2} = a_1 a_2^2 e^{a_2 c_1} > 0,$$

respectively. Thus the function g is convex and attains its minimum at

$$c_1^* = -\frac{1}{a_2} \log(a_1 a_2).$$

Further it holds that $g(0) = 0$ and $\lim_{c_1 \to \infty} g(c_1) = \infty$. Therefore the other root will be positive if and only if $c_1^* > 0$. Since a_2 is positive by assumption, the condition for a positive root can be written as $a_1 a_2 < 1$.
$\square$

PROPOSITION 2.6 *Let a_0, a_1, a_2, and a_3 be positive constants and $x \in$ $[a_2 - a_3, a_2]$. The solution of*

$$f(x) = a_0 + a_1 \left(\int_{y=0}^{a_2-x} f(x+y)\, dy + \int_{y=a_2-x}^{a_3} g(x+y)\, dy \right), \qquad (2.83)$$

is given by

$$
\begin{aligned}
f(x) = \; & a_1 \left(h(x) - h(a_2) \right) e^{-a_1 x} \\
& + \left(a_0 + a_1 \int_{y=0}^{a_3} g(a_2+y)\, dy \right) e^{a_1(a_2-x)},
\end{aligned}
\qquad (2.84)
$$

where

$$\frac{\partial h(x)}{\partial x} = g(x+a_3)\, e^{a_1 x}. \qquad (2.85)$$

PROOF OF PROPOSITION 2.6 *Define $\frac{\partial F(x)}{\partial x} = f(x)$ and $\frac{\partial G(x)}{\partial x} = g(x)$, then equation (2.83) can be written as*

$$\frac{\partial F(x)}{\partial x} = a_0 + a_1 \left(F(a_2) - F(x) + G(x+a_3) - G(a_2) \right). \qquad (2.86)$$

Differentiating both sides to x gives

$$\frac{\partial^2 F(x)}{\partial x^2} = -a_1 \left(\frac{\partial F(x)}{\partial x} - \frac{\partial G(x+a_3)}{\partial x} \right). \qquad (2.87)$$

From (2.87) we derive that

$$\frac{\partial F(x)}{\partial x} = \left(\int a_1 \frac{\partial G(x+a_3)}{\partial x} e^{a_1 x}\, dx + c \right) e^{-a_1 x}, \qquad (2.88)$$

where c is a constant to be determined. Thus

$$F(x) = \int \left(\int a_1 \frac{\partial G(x+a_3)}{\partial x} e^{a_1 x}\, dx + c \right) e^{-a_1 x}\, dx. \qquad (2.89)$$

Using the definition of h (see equation (2.85)) and integrating by parts gives

$$F(x) = - \left(h(x) + \frac{c}{a_1} \right) e^{-a_1 x} + G(x+a_3). \qquad (2.90)$$

Substitution of equation (2.90) in equation (2.86) gives

$$c = \left(a_0 + a_1 \int_{y=0}^{a_3} g\left(a_2 + y\right) dy \right) e^{a_1 a_2} - a_1 h\left(a_2\right). \qquad (2.91)$$

Equation (2.84) is found by substituting equation (2.91) in equation (2.88) and rearranging. □

PROPOSITION 2.7 *Let a_0, a_1, a_2, a_3, and a_4 be positive constants and $x \in [0, a_4]$. Then the solution of*

$$\frac{\partial f\left(x\right)}{\partial x} = a_0 + a_1 e^{-a_2 x} + a_3 \left(f\left(a_4\right) - f\left(x\right)\right), \qquad (2.92)$$

is given by

$$f\left(x\right) = \left(\frac{a_1}{a_2 - a_3} e^{-a_2 a_4} - \frac{a_0}{a_3} \right) e^{a_3 (a_4 - x)} - \frac{a_1}{a_2 - a_3} e^{-a_2 x} + c, \qquad (2.93)$$

where c is a constant to be determined by some boundary condition.

PROOF OF PROPOSITION 2.7 *Differentiating both sides of (2.92) to x gives*

$$\frac{\partial^2 f\left(x\right)}{\partial x^2} = -a_2 a_1 e^{-a_2 x} - a_3 \frac{\partial f\left(x\right)}{\partial x}. \qquad (2.94)$$

Solving this differential equation gives

$$\frac{\partial f\left(x\right)}{\partial x} = \delta_0 e^{-a_3 x} + \delta_1 e^{-a_2 x}. \qquad (2.95)$$

Substitution of (2.95) in (2.94) gives

$$\delta_1 = \frac{a_1 a_2}{a_2 - a_3}. \qquad (2.96)$$

Integrating equation (2.95) after substitution of equation (2.96) gives the following expression for f :

$$f\left(x\right) = -\frac{\delta_0}{a_3} e^{-a_3 x} - \frac{a_1}{a_2 - a_3} e^{-a_2 x} + c. \qquad (2.97)$$

Substitution of (2.95) and (2.97) in (2.92) gives

$$\delta_0 = \left(a_0 - \frac{a_1 a_3}{a_2 - a_3} e^{-a_2 a_4} \right) e^{a_3 a_4}. \qquad (2.98)$$

Equation (2.93) is found by substituting equation (2.98) in equation (2.97). □

PROPOSITION 2.8 *Let a_0, a_1, a_2 and a_3 be positive constants. Then the solution of*

$$f(x) = a_0 + a_1 \int_{y=0}^{a_2 - x} f(x + y) a_3 e^{-a_3 y} dy$$

$$+ a_1 \int_{y=a_2-x}^{\infty} g(x + y) a_3 e^{-a_3 y} dy, \qquad (2.99)$$

is given by

$$f(x) = \left(a_1 e^{a_2 a_3} \int_{y=a_2}^{\infty} g(y) a_3 e^{-a_3 y} dy - \frac{a_0 a_1}{1 - a_1} \right) e^{(1-a_1)a_3(x-a_2)}$$

$$+ \frac{a_0}{1 - a_1}. \qquad (2.100)$$

PROOF OF PROPOSITION 2.8 *Equation (2.99) can be written as*

$$f(x) = a_0 + a_1 e^{a_3 x} \left(\int_{y=x}^{a_2} f(y) a_3 e^{-a_3 y} dy + \int_{y=a_2}^{\infty} g(y) a_3 e^{-a_3 y} dy \right).$$

$$(2.101)$$

Define $\frac{\partial F(x)}{\partial x} = f(x) a_3 e^{-a_3 x}$, then (2.101) becomes

$$\frac{\partial F(x)}{\partial x} \frac{1}{a_3} e^{a_3 x} = a_0 + a_1 e^{a_3 x} (F(a_2) - F(x) + a_4), \qquad (2.102)$$

where

$$a_4 = \int_{y=a_2}^{\infty} g(y) a_3 e^{-a_3 y} dy. \qquad (2.103)$$

So, F must satisfy the following differential equation

$$\frac{\partial F(x)}{\partial x} = a_1 a_3 a_4 + a_0 a_3 e^{-a_3 x} + a_1 a_3 (F(a_2) - F(x)). \qquad (2.104)$$

Applying Proposition 2.7 to equation (2.104) gives

$$F(x) = \left(\frac{a_0}{1 - a_1} e^{-a_3 a_2} - a_4 \right) e^{a_1 a_3 (a_2 - x)} - \frac{a_0}{1 - a_1} e^{-a_3 x} + c. \qquad (2.105)$$

Thus

$$\frac{\partial F\left(x\right)}{\partial x} = \left(a_1 a_3 a_4 - \frac{a_0 a_1 a_3}{1-a_1}e^{-a_3 a_2}\right)e^{a_1 a_3(a_2-x)} + \frac{a_0 a_3}{1-a_1}e^{-a_3 x}.$$

(2.106)

From the definition of F it follows that

$$f\left(x\right) = \frac{1}{a_3}e^{a_3 x}\frac{\partial F\left(x\right)}{\partial x}.$$

(2.107)

Substitution of (2.106) into equation (2.107) gives equation (2.100). □

C.   DEFINITIONS
C.1   PROBABILITY DISTRIBUTIONS

DEFINITION 2.3 *The variable X is distributed according to a Poisson distribution with parameter $\mu > 0$ on $\mathbf{N}_0$ if the probability function, p_k, of X is given by*

$$p_k = \Pr\left(X = k\right) = \frac{\mu^k}{k!}e^{-\mu}, \quad for \ k = 0, 1, 2, \dots.$$

(2.108)

DEFINITION 2.4 *The variable X is distributed according to an exponential distribution with parameter $\mu > 0$ on the interval $(0, \infty)$ if the probability density function, $p\left(x\right)$, of X is given by*

$$p\left(x\right) = \mu e^{-\mu x}, \quad for \ x > 0.$$

(2.109)

DEFINITION 2.5 *The variable X is distributed according to a gamma distribution with parameters $\mu > 0$ and $n > 0$ on the interval $(0, \infty)$ if the probability density function, $p\left(x\right)$, of X is given by*

$$p\left(x\right) = \frac{\mu^n}{\Gamma\left(n\right)}x^{n-1}e^{-\mu x}, \quad for \ x > 0,$$

(2.110)

where $\Gamma\left(n\right)$ for $n > 0$ is defined by

$$\Gamma\left(n\right) = \int_{t=0}^{\infty} t^{n-1}e^{-t}dt.$$

(2.111)

Thus for $n \in \mathbf{N}$ we have that $\Gamma\left(n\right) = (n-1)!$.

DEFINITION 2.6 *The variable X is distributed according to a normal distribution with parameters $\mu \in \mathbf{R}$ and $\sigma > 0$ on the interval $(-\infty, \infty)$ if the probability density function, $p(x)$, of X is given by*

$$p(x) = \frac{1}{\sigma\sqrt{2\pi}}e^{-\frac{(x-\mu)^2}{2\sigma^2}}, \quad for\ x \in \mathbf{R}. \tag{2.112}$$

DEFINITION 2.7 *The variable X is distributed according to a uniform distribution on the interval $[a, b]$ if the probability density function, $p(x)$, of X is given by*

$$p(x) = \frac{1}{b-a}, \quad for\ x \in \mathbf{R}. \tag{2.113}$$

C.2   SETS

DEFINITION 2.8 $\mathbf{N}$ *is the set of natural numbers without zero, thus*

$$\mathbf{N} = \{1, 2, 3, \dots\}.$$

DEFINITION 2.9 $\mathbf{N}_0$ *is the set of natural numbers including zero, thus*

$$\mathbf{N}_0 = \{0, 1, 2, \dots\}.$$

DEFINITION 2.10 $\mathbf{R}$ *is the set of real numbers.*

DEFINITION 2.11 $\mathbf{R}_+$ *is the set of non-negative real numbers, thus*

$$\mathbf{R}_+ = \{x \in \mathbf{R} \,|\, x \geq 0\}.$$

C.3   OTHER

DEFINITION 2.12 $o(dt)$ *denotes a quantity which tends faster to zero than dt, i.e.*

$$\lim_{dt \downarrow 0} \frac{o(dt)}{dt} = 0.$$

D.   LEMMAS AND PROOFS

PROOF OF PROPOSITION 2.1 *From the analysis of Subsection A.2 we know that the threshold is unique if the following function is decreasing in θ*

$$\pi(\zeta_0) - (r+\lambda)(V(\theta) - I) + \lambda \int_{z \in S} (V(\theta + z) - I)\phi(z)\,dz, \tag{2.114}$$

where ϕ, with $\phi : S \rightarrow [0,1]$ and $S \subseteq \mathbf{R}_+$, is the probability density function of the probability distribution of the jump size. Due to the concavity of π and since ϕ is a probability density function we have

$$\int_{z \in S} \frac{\partial \pi \left(\theta + z \right)}{\partial \theta} \phi \left(z \right) dz \leq \frac{\partial \pi \left(\theta \right)}{\partial \theta}. \qquad (2.115)$$

Differentiating equation (2.114) with respect to θ and substitution of equation (2.115) into the result gives

$$- \left(\frac{r + \lambda}{r} \right) \frac{\partial \pi \left(\theta \right)}{\partial \theta} + \frac{\lambda}{r} \int_{z \in S} \frac{\partial \pi \left(\theta + z \right)}{\partial \theta} \phi \left(z \right) dz$$

$$\leq \; - \left(\frac{r + \lambda}{r} \right) \frac{\partial \pi \left(\theta \right)}{\partial \theta} + \frac{\lambda}{r} \frac{\partial \pi \left(\theta \right)}{\partial \theta}$$

$$= \; - \frac{\partial \pi \left(\theta \right)}{\partial \theta} < 0.$$

Thus equation (2.114) is decreasing in θ and the threshold is unique. $\square$

LEMMA 2.2 *Let X be a stochastic variable that is distributed over the interval $[0, \infty)$ according to some distribution with distribution function $F \left(x \right) = \Pr \left(X \leq x \right)$. Let $f \left(x \right)$ be a continuous and differentiable function on $[0, \infty)$. Then*

$$E \left[f \left(X \right) \right] = \int_{t=0}^{\infty} \frac{\partial f \left(t \right)}{\partial t} \left(1 - F \left(t \right) \right) dt + f \left(0 \right). \qquad (2.116)$$

PROOF OF LEMMA 2.2 *The proof is straightforward:*

$$E \left[f \left(X \right) \right] \;=\; \int_{x=0}^{\infty} f \left(x \right) dF \left(x \right) = \int_{x=0}^{\infty} \left[\int_{t=0}^{x} \frac{\partial f \left(t \right)}{\partial t} dt + f \left(0 \right) \right] dF \left(x \right)$$

$$=\; \int_{t=0}^{\infty} \left[\int_{x=t}^{\infty} \frac{\partial f \left(t \right)}{\partial t} dF \left(x \right) \right] dt + f \left(0 \right) \int_{x=0}^{\infty} dF \left(x \right)$$

$$=\; \int_{t=0}^{\infty} \frac{\partial f \left(t \right)}{\partial t} \left(1 - F \left(t \right) \right) dt + f \left(0 \right).$$

$\square$

LEMMA 2.3 *The discounted value of one unit of money the firm receives after the n-th technology arrival from now equals*

$$\left(\frac{\lambda}{r+\lambda}\right)^n. \tag{2.117}$$

PROOF OF LEMMA 2.3 *We want to derive an expression for*

$$E\left[e^{-r\left(T_{N(t)+n}-t\right)}\right]. \tag{2.118}$$

Since the technologies arrive according to a Poisson process, the time between two technology arrivals is exponentially distributed. We denote the number of technologies that arrive over an interval $[t, t+s)$ *by* $R(s)$. *Thus, it holds that* $R(s) = N(t+s) - N(t)$. *Due to the fact that* N *is a Poisson process with rate* λ, *the stochastic variable* $R(s)$ *is distributed according to a Poisson distribution with parameter* λs. *Now it is not hard to see that*

$$\Pr\left(T_{N(t)+n} - t \le s\right) = \sum_{k=n}^{\infty} \Pr\left(R(s) = k\right). \tag{2.119}$$

Using Lemma 2.2 we can derive the following expression:

$$E\left[e^{-r\left(T_{N(t)+n}-t\right)}\right] = 1 - r \int_{s=0}^{\infty} e^{-rs}\left(1 - \Pr\left(T_{N(t)+n} - t \le s\right)\right) ds. \tag{2.120}$$

Substitution of equation (2.119) into equation (2.120) gives

$$1 - r \int_{s=0}^{\infty} e^{-rs}\left(1 - \sum_{k=n}^{\infty}\Pr\left(R(s) = k\right)\right) ds$$

$$= 1 - r \int_{s=0}^{\infty} e^{-rs}\sum_{k=0}^{n-1}\Pr\left(R(s) = k\right) ds. \tag{2.121}$$

Since $R(s)$ *is Poisson distributed with parameter* λs *we have*

$$E\left[e^{-r\left(T_{N(t)+n}-t\right)}\right]$$

$$= 1 - r \sum_{k=0}^{n-1} \int_{s=0}^{\infty} e^{-rs} e^{-\lambda s}\frac{(\lambda s)^k}{k!} ds$$

$$= 1 - r \sum_{k=0}^{n-1} \frac{\lambda^k}{(r+\lambda)^{k+1}} \int_{s=0}^{\infty} \frac{(r+\lambda)^{k+1} e^{-(r+\lambda)s} s^k}{k!} ds. \tag{2.122}$$

The integral equals one, since it equals the cumulative distribution of a gamma distributed variable over its entire domain. Using the fact that

$$\sum_{i=0}^{j-1} a^i = \frac{1-a^j}{1-a},$$ (2.123)

gives

$$E\left[e^{-r(T_{N(t)+n}-t)}\right] = 1 - \frac{r}{(r+\lambda)}\frac{1-\left(\frac{\lambda}{r+\lambda}\right)^n}{1-\left(\frac{\lambda}{r+\lambda}\right)}$$

$$= \left(\frac{\lambda}{r+\lambda}\right)^n.$$ (2.124)

Thus equation (2.117) holds.   □

PROOF OF PROPOSITION 2.2 *Use Lemma 2.2 to calculate the expected value of t^* :*

$$E[t^*] = \int_{t=0}^{\infty} (1-\Pr(t^* \le t))\, dt = \int_{t=0}^{\infty} \left(1 - \sum_{n=n^*}^{\infty} \Pr(N(t)=n)\right) dt$$

$$= \int_{t=0}^{\infty} \sum_{n=0}^{n^*-1} \Pr(N(t)=n)\, dt = \int_{t=0}^{\infty} \sum_{n=0}^{n^*-1} e^{-\lambda t}\frac{(\lambda t)^n}{n!}\, dt$$

$$= \sum_{n=0}^{n^*-1} \frac{1}{\lambda} \int_{t=0}^{\infty} \frac{\lambda^{n+1} t^n}{n!} e^{-\lambda t}\, dt = \frac{n^*}{\lambda}.$$

In the second step we substitute equation (2.24) and in the fourth step the probability function of Poisson distribution (see Definition 2.3). The last equality holds, because the expression that is integrated is the density function of a gamma distribution with parameters λ and $n+1$ (see also Definition 2.5).

In the same way we derive the following expression for $E\left[(t^*)^2\right]$:

$$E\left[(t^*)^2\right] = \frac{n^*(n^*+1)}{\lambda^2}.$$ (2.125)

Thus

$$Var[t^*] = E\left[(t^*)^2\right] - (E[t^*])^2 = \frac{n^*}{\lambda^2}.$$

Thereby the proposition is proved.   □

PROOF OF PROPOSITION 2.3 *For this proof we use Lemma 2.2:*

$$E[t^*] = \int_{t=0}^{\infty} (1 - \Pr(t^* \le t)) \, dt = \int_{t=0}^{\infty} (1 - \Pr(\theta(t) \ge \theta^*)) \, dt$$

$$= \int_{t=0}^{\infty} \Pr(\theta(t) < \theta^*) \, dt$$

$$= \int_{t=0}^{\infty} \Pr\left(\theta_0 + \sum_{n=0}^{N(t)} u_n < \theta^*\right) dt$$

$$= \int_{t=0}^{\infty} \sum_{k=0}^{\infty} \Pr\left(\sum_{n=0}^{N(t)} u_n < \theta^* - \theta_0 \,\middle|\, N(t) = k\right) \Pr(N(t) = k) \, dt.$$

Noting that the sum of k independent and identically exponentially distributed variables (with parameter μ) is distributed according to a gamma distribution with parameters μ and k (see Definition 2.5) and that the probability that u_0 is less than $\theta^* - \theta_0$ is one, gives

$$E[t^*] = \int_{t=0}^{\infty} \left(e^{-\lambda t} + \sum_{k=1}^{\infty} \int_{x=0}^{\theta^*-\theta_0} \frac{\mu^k e^{-\mu x} x^{k-1}}{(k-1)!} \, dx \, \frac{(\lambda t)^k}{k!} e^{-\lambda t}\right) dt.$$

Rewriting gives

$$E[t^*] = \int_{x=0}^{\theta^*-\theta_0} e^{-\mu x} \sum_{k=1}^{\infty} \frac{\mu^k x^{k-1}}{\lambda(k-1)!} \int_{t=0}^{\infty} \frac{\lambda^{k+1} t^k}{k!} e^{-\lambda t} \, dt \, dx + \frac{1}{\lambda}.$$

Seeing that the second integral equals one (probability density function of gamma distribution) and knowing that $\sum_{k=1}^{\infty} \frac{(\mu x)^{k-1}}{(k-1)!} = e^{\mu x}$ gives

$$E[t^*] = \frac{\mu}{\lambda} \int_{x=0}^{\theta^*-\theta_0} dx + \frac{1}{\lambda} = \frac{\mu(\theta^* - \theta_0) + 1}{\lambda}.$$

In the same way it can be derived that

$$E\left[(t^*)^2\right] = \frac{2 + 4\mu(\theta^* - \theta_0) + \mu^2(\theta^* - \theta_0)^2}{\lambda^2}.$$

So

$$Var\left[t^*\right] = \frac{1 + 2\mu\left(\theta^* - \theta_0\right)}{\lambda^2}.$$

□

Chapter 3

DECREASING INVESTMENT COST

1.   INTRODUCTION

We consider a firm whose profit is only influenced by its own technology choice. There are two differences with the model of Chapter 2. First, it is assumed that the efficiency improvements of the new technologies are known. In practice this does not seem to be a very restrictive assumption. For example, when Intel launched the Pentium processor everyone knew that one day they would come up with a processor that is twice as fast as the Pentium processor. The only thing not known for sure was when this processor would become available. Second, the prices of new technologies are assumed to drop over time, implying that a firm needs to invest less in case it decides to buy a new technology at a later point of time. The reason for this price decrease is that, as time passes, the demand for a particular technology declines because of market saturation and the invention of newer technologies that are better than this particular one.

The problem of the firm is (1) to decide whether to invest in a new technology or not, (2) if the firm decides to invest, which technology to choose, and (3) at what time it is optimal to invest. The sooner the firm invests the higher the price it has to pay for a new technology, but the sooner the firm can produce more efficiently. Another disadvantage of investing very fast is that there exists a risk that a much better technology will become available just a little later. Both the single switch and multiple switch case will be analyzed.

The model will be solved using dynamic programming. Similar to the previous chapter we compare the optimal investment strategy with the one that would have been found when the widely used net present

51

value method was applied. The net present value method prescribes that the firm should go ahead with investing as soon as the discounted cash flow stream exceeds the initial sunk cost investment. In doing so the net present value method does not take into account the advantage of delaying investment, which arises from the fact that the later a firm invests in a new technology the lower the sunk cost investment will be. Also, this method does not take into account the probability that a better technology will be invented at a later point of time. Hence, applying the net present value rule will lead to a suboptimal outcome, because the option value of postponing the investment is not taken into account. In this respect this model contributes to a recent stream of literature in which investment decisions are analyzed as real options (for an overview see the well received book by Dixit and Pindyck, 1996).

This chapter is organized as follows. In Section 2 we present the basic model, while in Section 3 the optimal investment strategy for the single switch case is derived. The multiple switch case is discussed in Section 4. We compare the optimal strategy with the one that is the result of applying the net present value rule in Section 5. Section 6 concludes.

2.   THE MODEL

Consider a risk-neutral firm whose profit is not influenced by the technology choice of other firms. Since the firm can make more profits with a more efficient technology we assume that the firm has a profit function π which is increasing and concave in $\zeta \, (\geq 0)$, the technology-efficiency parameter. We analyze a dynamic model with an infinite planning period and assume that the firm maximizes its value and discounts against rate $r \, (> 0)$. Initially, at time $t = 0$, the firm produces with a technology designated by $\zeta \, (0) = \zeta_0$, with $\zeta_0 \geq 0$. As time passes new technologies become available, and the firm has the opportunity to adopt a new technology. At time $t \, (\geq 0)$ the efficiency of the most efficient available technology is denoted by $\theta \, (t)$ and the efficiency of the technology that the firm uses is denoted by $\zeta \, (t)$. We assume that the process of technological evolution is exogenous to the firm. Technologies become more and more efficient over time, and the more efficient a technology the larger the associated parameter θ. Thus $0 \leq \zeta \, (t) \leq \theta \, (t)$ for all $t \geq 0$. However, the arrival process of the new technologies is a stochastic process. On the other hand, the efficiency improvements of the new technologies are assumed to be known. The i-th new technology has an (known) efficiency level equal to $\theta_i \, (> \theta_{i-1})$ with $i \in \mathbb{N}$ and $\theta_0 \geq \zeta_0$. In the remainder of this chapter we will use the efficiency level of a technology to refer to that specific technology. For example we write technology θ_i instead of technology i. We denote the number of technology arrivals over the interval $[0, t)$ by $N(t)$. Therefore $\theta(t) = \theta_{N(t)}$ for all $t \geq 0$. For notational convenience we write θ instead of $\theta(t)$. We assume that $N(t)$ is a Poisson process with rate $\lambda > 0$. We denote the time elapsed between the invention of technology $i - 1$ and technology i by τ_i. We define

$$T_i = \sum_{j=1}^{i} \tau_j,$$

thus T_i is equal to the point in time at which technology θ_i becomes available. Let $T(t)$, with $t \geq 0$, be the set which contains the arrival dates of the $N(t)$ technologies that arrived over the time interval $[0, t)$. Thus $T(t) = \emptyset$ if $N(t) = 0$ and $T(t) = \{T_1, \dots, T_{N(t)}\}$ if $N(t) \geq 1$. Further it holds that $|T(t)| = N(t)$, where $|S|$ is defined to be equal to the number of elements of a finite set S.

When the firm adopts technology θ_i at time t it incurs a non negative sunk cost investment which is denoted by $I_i(t)$, with $I_i : [T_i, \infty) \to \mathbf{R}_+$. At the moment that a new technology becomes available to the market, the investment cost instantaneously declines with a certain fraction, which is given a Poisson jump. Here the investment cost is subject to the same Poisson process as the one that determines the arrival rate of new technologies. The reason for this can be that the suppliers of technologies put the old technology for sale against a lower price to get rid of these less efficient technologies.

For $t \geq T_i$ the investment cost $I_i(t)$ decreases according to the following process

$$dI_i(t) = \begin{cases} -\alpha_i I_i(t)\, dt - \beta_i I_i(t) & \text{with probability } \lambda dt, \\ -\alpha_i I_i(t)\, dt & \text{with probability } 1 - \lambda dt, \end{cases} \quad (3.1)$$

$$I_i(T_i) = I_{i0}, \quad (3.2)$$

where $I_{i0} \geq 0$. α_i, with $\alpha_i \geq 0$, is the parameter that determines the speed of the deterministic decline of the investment cost and $\beta_i \in [0, 1)$ corresponds to the size of the jump the investment cost makes when a new technology arrives. The deterministic decline of the investment cost can be explained by market saturation: the price of a certain technology decreases over time due to the fact that the demand for that technology decreases.

The following proposition gives the solution of the system of equations (3.1)-(3.2). The proof is given in Appendix A.

PROPOSITION 3.1 *The investment cost of technology i at time $t \geq T_i$ is equal to*

$$I_i(t) = (1 - \beta_i)^{N(t)-i} I_{i0} e^{-\alpha_i(t-T_i)}. \tag{3.3}$$

The general problem facing the firm is: (1) to choose to which technology to switch and (2) to choose the right moment to switch to that technology.

3.   SINGLE SWITCH

In this section we assume that the firm is allowed to switch technologies only once. One reason that a firm cannot invest more than once can be that the firm's financial means are limited. There are an infinite number of possible investment strategies for the firm. The first strategy is "never invest" and the others are of the form: "invest in technology θ_i", $i \in \mathbb{N}$. Although there are an infinite number of investment strategies, only a finite number need to be considered. The reason is that the strategies "invest in technology θ_i" for i sufficiently large may be ignored. Thus, the values of these strategies are the same as the value of the strategy "never invest" and therefore they can be ignored for the moment. An equivalent application of this so-called forecast horizon procedure can be found in Nair, 1995.

We know that the firm is going to invest at some time anyway. The reason for this is a combination of (i) that the investment costs go to zero as time goes to infinity, (ii) that the efficiency levels of the new technologies are higher than the efficiency level of the technology that is currently in use, and (iii) the profit function is increasing and concave in the efficiency level of the technology in use.

We obtain the optimal investment strategy for the firm by comparing, for all values of the investment costs I_i, $i \in \mathbb{N}$, the value of the firm under the possible strategies. Therefore we consider that the firm is going to invest in some technology θ_i, with $i \in \mathbb{N}$, determine the optimal time to invest in this technology, and derive the value of the firm conditional on this investment strategy.

The expected value of the firm at time t, if the firm has not invested yet, is denoted by $F(t, T(t), \zeta_0)$ and is equal to

$$F(t, T(t), \zeta_0) = \max_{i \in \mathbb{N}} (F_i(t, T(t), \zeta_0)), \tag{3.4}$$

where $F_i(t, T(t), \zeta_0)$ equals the value of the firm at time t if the firm decides to invest in technology θ_i, given $T(t)$ and ζ_0. In order to derive an expression for $F_i(t, T(t), \zeta_0)$ we first consider the case where technology θ_i is already available at time t ($i \leq N(t)$) and after that the case where technology θ_i is not yet invented at time t ($i > N(t)$).

3.1   TECHNOLOGY BEING AVAILABLE

In this subsection we derive the value of the firm when for some $i \in \mathbf{N}$ the firm is going to invest in technology θ_i and technology θ_i has already been invented. The only thing that is left is to determine the timing of the investment. The problem facing the firm is an optimal stopping problem. For an introduction we refer the reader to Appendix A of Chapter 2. Intuition suggests that there will be a critical level I_i^* such that it is optimal for the firm to invest when the investment cost is equal or below the critical level, $I_i \leq I_i^*$, and it is optimal to wait with investing otherwise, $I_i > I_i^*$. The following proposition, that is proved in Appendix A, guarantees the uniqueness of the threshold.

PROPOSITION 3.2 *The threshold I_i^* is unique.*

There are two ways for the investment cost to fall below the critical level (see Figure 3.1). One possible way is by a jump and the other possibility is that the investment cost passes the critical level smoothly. Therefore we can identify three regions for the investment cost. In the first region the investment cost is above the critical level and it is not possible that the investment cost falls below the critical level after the next jump. The second region is characterized by the facts that the investment cost is above the critical level and that the investment cost will be below the critical level after the next jump. In the last region the investment cost is below the critical level. The first two regions together are called the continuation region. The boundary between the continuation region and the third region is of course the critical level I_i^*. The cut-off point between the first and the second region is determined by the critical level and the size of the jump, so that it equals $\frac{I_i^*}{(1-\beta_i)}$. The next step is to derive expressions for the value of the firm, denoted by $f_i(I_i, \zeta_0)$, in each of these three regions and an expression for the critical level.

In the termination region, $\{I_i \mid 0 \leq I_i \leq I_i^*\}$, it is optimal to invest right away in technology θ_i. Since the firm can make only one technology switch, the firm will produce with technology θ_i forever after the switch. Thus, in this region the value of the firm equals

$$f_i(I_i, \zeta_0) = \int_{s=0}^{\infty} \pi(\theta_i) e^{-rs} ds - I_i = \frac{\pi(\theta_i)}{r} - I_i. \qquad (3.5)$$

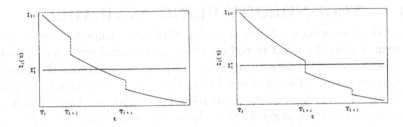

Figure 3.1. Sample paths of $I_i(t)$. In the left panel the investment cost decreases smoothly through the critical level I_i^* and in the right panel the investment cost jumps through the critical level I_i^*.

The value function f_i must satisfy the following Bellman equation in the continuation region (cf. Appendix A of Chapter 2):

$$r f_i(I_i, \zeta_0) = \pi(\zeta_0) + \lim_{dt \downarrow 0} \frac{1}{dt} E\left[df_i(I_i, \zeta_0)\right]. \tag{3.6}$$

In the second part of the continuation region, where $I_i \in \left(I_i^*, \frac{I_i^*}{(1-\beta_i)}\right)$, the firm is going to switch after the next jump of the investment cost. Since the firm can make only one technology switch, we know that the firm is going to produce with this technology θ_i forever after the technology switch. So the value of the firm after the technology switch equals $\frac{\pi(\theta_i)}{r}$. Expanding $E\left[df_i(I_i, \zeta_0)\right]$ using Itô's lemma (see Appendix A of Chapter 2) and equation (3.1) gives

$$
\begin{aligned}
E\left[df_i(I_i, \zeta_0)\right] = \ & (1 - \lambda dt)\left(-\frac{\partial f_i(I_i, \zeta_0)}{\partial I_i}\alpha_i I_i dt\right) \\
& + \lambda dt\left(\frac{\pi(\theta_i)}{r} - (I_i - \alpha_i I_i dt - \beta_i I_i) - f_i(I_i, \zeta_0)\right) \\
& + o(dt).
\end{aligned}
\tag{3.7}
$$

Substitution of equation (3.7) into (3.6) and rewriting gives

$$f_i(I_i, \zeta_0) + \frac{\partial f_i(I_i, \zeta_0)}{\partial I_i}\frac{\alpha_i}{r+\lambda}I_i + \frac{\lambda(1-\beta_i)}{r+\lambda}I_i = \frac{\pi(\zeta_0)}{r+\lambda} + \frac{\lambda}{r+\lambda}\frac{\pi(\theta_i)}{r}. \tag{3.8}$$

The solution of this differential equation is given by

$$f_i(I_i, \zeta_0) = B_i I_i^{-\frac{(r+\lambda)}{\alpha_i}} - \frac{\lambda(1-\beta_i)}{r+\alpha_i+\lambda}I_i + \frac{\pi(\zeta_0)}{r+\lambda} + \frac{\lambda}{r+\lambda}\frac{\pi(\theta_i)}{r}, \tag{3.9}$$

in which B_i is a constant to be determined. Hence, the value of the firm in the second region consists of four parts. The first part is equal to the value of the option to invest in technology θ_i. Note that this option value increases over time with rate α_i (not $\alpha_i + \lambda\beta_i$) since the firm kills this option after the next technology arrival. The latter is also the reason for the arrival rate λ being added to the discount rate (see also Subsection 3.1). The second part is the discounted value of the investment cost to be paid when adopting technology θ_i. To explain this, let S denote the time till the next technology arrival, so that

$$
\begin{aligned}
E\left[e^{-rS}I_i(S)\right] &= \int_{s=0}^{\infty} E\left[e^{-rS}I_i(S)\middle| S = s\right]\lambda e^{-\lambda s}ds \\
&= \int_{s=0}^{\infty} e^{-rs}\left(1 - \beta_i\right)I_i e^{-\alpha_i s}\lambda e^{-\lambda s}ds \\
&= \frac{\lambda\left(1 - \beta_i\right)}{r + \alpha_i + \lambda}I_i.
\end{aligned}
$$

The third part is the expected value of the discounted profit flow the firm generates from now until the next technology arrival. Therefore λ is added to the discount rate. The last part is the discounted value of the firm after the technology switch. A general derivation and interpretation of the discount factor $\frac{\lambda}{r+\lambda}$ is given by Lemma 2.3 in Appendix D of Chapter 2.

Expressions for B_i and the critical level I_i^* can be derived by considering the case that the investment cost decreases smoothly through the critical level. Then the value matching and smooth pasting conditions must hold at the critical level.

In the first region, i.e. $\left\{I_i \middle| \frac{I_i^*}{(1-\beta_i)} < I_i \leq I_{i0}\right\}$, we know that the firm is not going to switch after the next jump of the investment cost. In this case the function f_i must satisfy the following equation

$$
f_i\left(I_i, \zeta_0\right) + \frac{\partial f_i\left(I_i, \zeta_0\right)}{\partial I_i}\frac{\alpha_i + \lambda\beta_i}{r}I_i = \frac{\pi\left(\zeta_0\right)}{r}. \tag{3.10}
$$

The solution of this differential equation equals

$$
f_i\left(I_i, \zeta_0\right) = A_i I_i^{-\frac{r}{\alpha_i + \lambda\beta_i}} + \frac{\pi\left(\zeta_0\right)}{r}, \tag{3.11}
$$

where A_i is a constant to be determined later on. From (3.11) we obtain that the value of the firm in the first region consists of two parts. The first part can be looked upon as the value of the option to invest in technology θ_i. The second part is equal to the value of the firm if it decides to produce with technology ζ_0 forever, which then equals the value of the firm when the firm never exercises the option to invest. The constant A_i can be determined by making use of the continuity condition at the boundary between the first and the second region.

Now we can derive expressions for the constants and the cut-off points using the following three conditions. First the continuity condition at the cut-off point between the first two regions, which is derived from (3.11) and (3.9):

$$
B_i \left(\frac{I_i^*}{1 - \beta_i} \right)^{-\frac{r+\lambda}{\alpha_i}} - \frac{\lambda}{r + \alpha_i + \lambda} I_i^* + \frac{\pi(\zeta_0)}{r + \lambda} + \frac{\lambda}{r + \lambda} \frac{\pi(\theta_i)}{r}
$$
$$
= A_i \left(\frac{I_i^*}{1 - \beta_i} \right)^{-\frac{r}{\alpha_i + \lambda \beta_i}} + \frac{\pi(\zeta_0)}{r}. \tag{3.12}
$$

Second the value matching condition at the cut-off point between the second and the third region, stating that the firm at this cut-off point is indifferent between investing right away and waiting just a little bit longer before making the investment:

$$
B_i (I_i^*)^{-\frac{r+\lambda}{\alpha_i}} - \frac{\lambda(1 - \beta_i)}{r + \alpha_i + \lambda} I_i^* + \frac{\pi(\zeta_0)}{r + \lambda} + \frac{\lambda}{r + \lambda} \frac{\pi(\theta_i)}{r} = \frac{\pi(\theta_i)}{r} - I_i^*. \tag{3.13}
$$

The following smooth pasting condition holds at the cut-off point between the second and the third region,

$$
\frac{\partial}{\partial I_i} \left[B_i I_i^{-\frac{r+\lambda}{\alpha_i}} - \frac{\lambda(1 - \beta_i)}{r + \alpha_i + \lambda} I_i + \frac{\pi(\zeta_0)}{r + \lambda} + \frac{\lambda}{r + \lambda} \frac{\pi(\theta_i)}{r} \right] \bigg|_{I_i = I_i^*}
$$
$$
= \frac{\partial}{\partial I_i} \left[\frac{\pi(\theta_i)}{r} - I_i \right] \bigg|_{I_i = I_i^*}. \tag{3.14}
$$

Rewriting (3.14) gives the following expression for B_i:

$$
B_i = \frac{\alpha_i}{r + \lambda} \frac{r + \alpha_i + \lambda \beta_i}{r + \alpha_i + \lambda} I_i^{*\frac{(r+\lambda)}{\alpha_i} + 1}. \tag{3.15}
$$

Substitution of (3.15) in (3.13) and solving for I_i^* gives

$$
I_i^* = \frac{\pi(\theta_i) - \pi(\zeta_0)}{r + \alpha_i + \lambda \beta_i}. \tag{3.16}
$$

An economic interpretation of (3.16) is given in Subsection 3.4. Note that the critical level does only depend on technologies 0 and i. This is for the reason that the critical level is derived conditional on the fact that the firm is going to switch from technology ζ_0 to technology θ_i. By combining equations (3.12), (3.15) and (3.16) an expression for A_i can be derived.

We conclude that the value of the firm at time t conditional on the strategy "invest in technology θ_i" when technology θ_i is available at time t is given by

$$
f_i\left(I_i, \zeta_0\right) = \begin{cases} A_i\left(I_i\right)^{-\frac{r}{\alpha_i + \lambda \beta_i}} + \frac{\pi(\zeta_0)}{r} & \text{if } \frac{I_i^*}{(1-\beta_i)} < I_i \le I_{i0}, \\ B_i\left(I_i\right)^{-\frac{r+\lambda}{\alpha_i}} - \frac{\lambda(1-\beta_i)}{r+\alpha_i+\lambda} I_i \\ +\frac{\pi(\zeta_0)}{(r+\lambda)} + \frac{\lambda}{r+\lambda}\frac{\pi(\theta_i)}{r} & \text{if } I_i^* < I_i \le \frac{I_i^*}{(1-\beta_i)}, \\ \frac{\pi(\theta_i)}{r} - I_i & \text{if } 0 \le I_i \le I_i^*. \end{cases} \tag{3.17}
$$

Using this equation we get the following expression for $F_i\left(t, T\left(t\right), \zeta_0\right)$ with $t \ge T_i$:

$$
F_i\left(t, T\left(t\right), \zeta_0\right) = f_i\left(I_i\left(t\right), \zeta_0\right), \tag{3.18}
$$

where $I_i\left(t\right)$ is calculated using equation (3.3).

3.2   TECHNOLOGY NOT BEING AVAILABLE

If at time t technology θ_i is not yet available for the firm, the value of the firm under the strategy "invest in technology θ_i" is equal to the sum of the discounted profit flows generated on the time interval (t, T_i), where T_i is the moment of time that technology θ_i is invented, and the discounted value of the firm under the strategy "invest in technology θ_i" at time T_i, i.e.

$$
\begin{aligned}
F_i\left(t, T\left(t\right), \zeta_0\right) &= E\left[\int_{s=0}^{T_i-t} \pi\left(\zeta_0\right) e^{-rs} ds + F_i\left(T_i, T\left(T_i\right), \zeta_0\right) e^{-r(T_i-t)}\right] \\
&= \frac{\pi\left(\zeta_0\right)}{r} + \left(F_i\left(T_i, T\left(T_i\right), \zeta_0\right) - \frac{\pi\left(\zeta_0\right)}{r}\right) \\
&\quad \times E\left[e^{-r(T_i-t)}\right].
\end{aligned} \tag{3.19}
$$

From Lemma 2.3 in Appendix D of Chapter 2 we know that

$$
E\left[e^{-r(T_i-t)}\right] = \left(\frac{\lambda}{r+\lambda}\right)^{i-N(t)}. \tag{3.20}
$$

Substitution of the equations (3.18) and (3.20) in equation (3.19) yields for $t < T_i$,

$$
F_i\left(t, T\left(t\right), \zeta_0\right) = \left(\frac{\lambda}{r+\lambda}\right)^{i-N(t)} A_i\left(I_{i0}\right)^{-\frac{r}{\alpha_i + \lambda \beta_i}} + \frac{\pi(\zeta_0)}{r}, \tag{3.21}
$$

if $\frac{I_i^*}{(1-\beta_i)} < I_{i0}$,

$$F_i\left(t, T\left(t\right), \zeta_0\right) = \left(\frac{\lambda}{r+\lambda}\right)^{i-N(t)} \left(B_i\left(I_{i0}\right)^{-\frac{r+\lambda}{\alpha_i}} - \frac{\lambda(1-\beta_i)}{r+\alpha_i+\lambda} I_{i0}\right)$$

$$+ \left(\frac{\lambda}{r+\lambda}\right)^{i-N(t)+1} \left(\frac{\pi(\theta_i)}{r} - \frac{\pi(\zeta_0)}{r}\right) + \frac{\pi(\zeta_0)}{r}, \qquad (3.22)$$

if $I_i^* < I_{i0} \leq \frac{I_i^*}{(1-\beta_i)}$, and

$$F_i\left(t, T\left(t\right), \zeta_0\right) = \left(\frac{\lambda}{r+\lambda}\right)^{i-N(t)} \left(\frac{\pi(\theta_i)}{r} - I_{i0}\right)$$

$$+ \left(1 - \left(\frac{\lambda}{r+\lambda}\right)^{i-N(t)}\right) \frac{\pi(\zeta_0)}{r}, \qquad (3.23)$$

if $0 \leq I_{i0} \leq I_i^*$.

3.3   OPTIMAL INVESTMENT STRATEGY

From the above analysis it can be concluded that the following theorem states the optimal investment strategy.

THEOREM 3.1 *Consider a time $t\ (\geq 0)$. Let $k = \arg\max_{i\in\mathbf{N}} F_i\left(t, T\left(t\right), \zeta_0\right)$. Given that the firm has not invested before, it is optimal for the firm to invest in technology θ_k at time t if the following two conditions are fulfilled:*

(1)	$I_k\left(t\right) < I_k^*$,	(3.24)
(2)	$k \leq N\left(t\right)$,	(3.25)

and otherwise it is optimal to wait with investing. The value functions $F_i\left(t, T\left(t\right), \zeta_0\right)$ for $i \in \{1, \ldots, N\left(t\right)\}$ and for $i > N\left(t\right)$ are calculated with equations (3.18) and (3.21)-(3.23), respectively. The investment costs $I_i\left(t\right)$, $i \in \{1, \ldots, N\left(t\right)\}$, are calculated with equation (3.3) and $I_i\left(t\right) = I_{i0}$ for $i > N\left(t\right)$. The optimal switching levels I_i^*, for $i \in \mathbf{N}$, are given by equation (3.16).

The first condition ensures that the sunk cost investment is already below the critical value so that it is optimal to invest right now in θ_k. The second condition says that technology θ_k is already available. If at least one of these conditions is not fulfilled, it is optimal for the firm to keep on producing with the old technology and wait with investing.

As mentioned before, only a finite number of technologies need to be considered. The reason is that the strategies "invest in technology

θ_i" for i sufficiently large may be ignored. Since $\lim_{i\to\infty}\left(\frac{\lambda}{r+\lambda}\right)^{i-N(t)}=0$, the value of the firm under the strategy "invest in technology θ_i" for sufficiently large i will equal $\frac{\pi(\zeta_0)}{r}$ (let i go to infinity in equations (3.21)-(3.23)). Thus, the values of these strategies are the same as the value of the strategy "never invest" and therefore they can be ignored for the moment.

EXAMPLE 3.1 *In this example we again analyze the firm from Example 2.1. Thus the profit flow is equal to $\pi(\zeta)=200\zeta^2$ and the discount rate is given by $r=0.1$. We assume that on average each five years a new technology arrives, i.e. $\lambda=\frac{1}{5}$, and that the efficiency of the i-th technology is given by $\theta_i=\theta_0+\frac{1}{2}i$. Further we set $\zeta_0=\theta_0=1$. The parameters for the investment cost are the same for all new technologies and equal to $I_{i0}=I_0=1600$, $\alpha_i=\alpha\in[0,0.9]$ and $\beta_i=0$, for all $i\in\mathbb{N}$. We consider an interval of α values rather than a unique number, because we want to analyze the effect of α on the optimal investment strategy of the firm.*

For the parameter values concerned it holds that $I_4^*\geq I_0$, which implies that it is optimal to adopt technology θ_4 at its arrival date. Consider the decision problem of the firm at time T_4. Investing in technology θ_4 gives a payoff of

$$\frac{\pi(\theta_4)}{r}-I_0=16400.$$

When the firm waits for the next technology its expected value equals

$$\frac{\pi(\zeta_0)}{r}+\frac{\lambda}{r+\lambda}\left(\frac{\pi(\theta_5)-\pi(\zeta_0)}{r}-I_0\right)=15933.$$

It turns out that the expected value of the firm is even lower when it waits for better technologies than technology θ_5. Therefore the optimal strategy at time T_4 is to adopt technology θ_4.

Further it can be obtained that technologies θ_1 and θ_2 will never be adopted by the firm, because waiting for technology θ_4 yields (in any case) a higher payoff:

$$4500=\frac{\pi(\theta_1)}{r}<\frac{\pi(\zeta_0)}{r}+\left(\frac{\lambda}{r+\lambda}\right)^3\left(\frac{\pi(\theta_4)-\pi(\zeta_0)}{r}-I_0\right)=6267,$$

and

$$8000=\frac{\pi(\theta_2)}{r}<\frac{\pi(\zeta_0)}{r}+\left(\frac{\lambda}{r+\lambda}\right)^2\left(\frac{\pi(\theta_4)-\pi(\zeta_0)}{r}-I_0\right)=8400.$$

Adopting technology θ_3 can be optimal, since

$$12500 = \frac{\pi(\theta_3)}{r} > \frac{\pi(\zeta_0)}{r} + \frac{\lambda}{r+\lambda}\left(\frac{\pi(\theta_4) - \pi(\zeta_0)}{r} - I_0\right) = 11600.$$

Consider the case that technology θ_4 is not yet available. Then adopting technology θ_3 is optimal when

$$\frac{\pi(\theta_3)}{r} - I_3(t) \geq 11600,$$

i.e. for $t \geq t_3^*$, where

$$t_3^* = \frac{1}{\alpha}\log\left(\frac{16}{9}\right).$$

Thus the probability that the firm adopts technology θ_3, given α, equals

$$\Pr(\theta_3|\,\alpha) = \Pr(T_4 - T_3 > t_3^*) = 1 - \Pr(T_4 - T_3 \leq t_3^*) = e^{-\frac{t_3^*}{5}}.$$

With the complementary probability technology θ_4 is adopted:

$$\Pr(\theta_4|\,a) = 1 - e^{-\frac{t_3^*}{5}}.$$

These probabilities are plotted in Figure 3.2. Notice that when $\alpha = 0$ we are back in the analysis of Chapter 2 and the firm is going to adopt technology θ_4 for sure. Increasing α increases the probability that technology θ_3 is adopted. This for the reason that increasing α increases the probability that the investment cost of technology θ_3 has decreased enough to make its adoption optimal.

3.4   COMPARATIVE STATICS

By partially differentiating equation (3.16) we can show that the critical switching level I_i^* will be higher, implying that the firm is willing to pay more for technology θ_i, the smaller the discount rate, the higher the efficiency level of the new technology, and the lower the efficiency level of the technology that is currently used. The intuition for these effects is straightforward: the opportunity cost of waiting in anticipation of a lower investment cost consists of the discounted forgone profits during the waiting period, which clearly will be greater the smaller is r. A relatively higher efficiency level of the new technology, or a relatively lower efficiency level of the technology currently in use, makes a technology switch more attractive. Therefore the optimal switching level will be higher, which implies an earlier technology adoption.

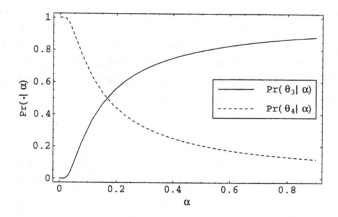

Figure 3.2. Probability of adopting technology θ_3 and probability of adopting technology θ_4, both as function of α.

Further we see that the critical switching level I_i^* will be lower, the higher α_i, the higher λ and the higher β_i. This means that the firm anticipates to a more rapid decline of the sunk cost investment by waiting for a smaller sunk investment cost.

4.   MULTIPLE SWITCHES

In this section we try to extend the analysis of the previous section to the case where the firm is allowed to make n technology switches. Let $G_n\left(t, T\left(t\right), \zeta_{n-1}\right)$ denote the value of the firm before the n-th technology switch at time t when the firm produces with technology ζ_{n-1} and the arrival dates of new technologies are summarized in $T\left(t\right)$. Then we know that

$$G_n\left(t, T\left(t\right), \zeta_{n-1}\right) = F\left(t, T\left(t\right), \zeta_{n-1}\right), \qquad (3.26)$$

where F is defined by equation (3.4). In the same fashion we define the value of the firm before the i-th technology switch, with $i \in \{1, \ldots, n-1\}$:

$$G_i\left(t, T\left(t\right), \zeta_{i-1}\right) = \max_{j \in \{k | \theta_k > \zeta_{i-1}\}} G_{ij}\left(t, T\left(t\right), \zeta_{i-1}\right). \qquad (3.27)$$

In the last equation G_{ij} is the expected value if the firm makes its i-th technology switch to technology θ_j. Assume for the moment that technology θ_j is available at time t, $j \leq N\left(t\right)$, and that there exists a unique threshold I_{ij}^*. Hence, adopting technology θ_j is optimal if the investment cost is equal or below the threshold and waiting is optimal when the investment cost is larger than the threshold. By $g_{ij}(I_j, T(t), \zeta_{i-1})$ we denote the value of the firm when ζ_{i-1} is the current technology in use, the firm is going to adopt technology j, which is available, but the investment cost I_j is currently above the critical level I_{ij}^* and $T(t)$ is the set with arrival dates. Then g_{ij} must satisfy the following Bellman equation

$$r g_{ij}(I_j, T(t), \zeta_{i-1}) = \pi(\zeta_{i-1}) + \lim_{dt \downarrow 0} \frac{1}{dt} E[dg_{ij}(I_j, T(t), \zeta_{i-1})]. \quad (3.28)$$

Expanding $E[dg_{ij}]$, given that adopting technology θ_j is not optimal after the next technology arrival, i.e. $(1 - \beta_j) I_j \geq I_{ij}^*$, with Itô's lemma gives

$$
\begin{aligned}
E&[dg_{ij}(I_j, T(t), \zeta_{i-1})] \\
&= \lambda dt \left(g_{ij}((1 - \beta_j) I_j, T(t) \cup \{t + dt\}) - g_{ij}(I_j, T(t), \zeta_{i-1}) \right) \\
&\quad + (1 - \lambda dt) \left(-\frac{\partial g_{ij}(I_j, T(t), \zeta_{i-1})}{\partial I_j} \alpha_j I_j dt \right) + o(dt). \quad (3.29)
\end{aligned}
$$

Substitution of equation (3.29) into equation (3.28) gives

$$
\begin{aligned}
(r + \lambda) g_{ij}(I_j, T(t), \zeta_{i-1}) &= \pi(\zeta_{i-1}) \\
&\quad + \lambda g_{ij}((1 - \beta_j) I_j, T(t) \cup \{t\}, \zeta_{i-1}) \\
&\quad - \frac{\partial g_{ij}(I_j, T(t), \zeta_{i-1})}{\partial I_j} \alpha_j I_j. \quad (3.30)
\end{aligned}
$$

Following the steps of the previous section we first have to solve the last equation for g_{ij}. However, it is impossible to derive a closed form expression for g_{ij} from equation (3.30). The problem is mainly caused by the fact that in the model of this chapter, contrary to the model of Chapter 2, at time t we can not ignore the technologies with efficiencies in the interval $(\zeta(t), \theta(t))$. In the previous chapter a technology was either adopted at its arrival date or not at all. However, in this chapter we have to take into account all technologies, since it is possible that due to a late arrival of another technology it is optimal to adopt an already existing technology (cf. Example 3.1). Further in the multiple switch case the firm's value under the strategy "the following technology to adopt is technology θ_j" increases not only as a result of the decrease of the investment cost of technology θ_j, but also through the decreases in the investment costs of technologies $\theta_{j+1}, \dots, \theta_{N(t)}$ and the arrival of new technologies. Since we need an explicit equation for g_{ij} to be able to compare the multiple and single switch cases, we conclude that the multiple switch case can, unfortunately, not be explicitly solved. Though, it may be possible that equation (3.30) can be solved numerically.

5. NET PRESENT VALUE METHOD

In this section we derive the investment strategy according to the net present value method and compare the result with the optimal investment strategy derived in the previous sections.

Consider the case in which the firm adopts technology θ_i. If at time, say $t = t_0$, investment in technology θ_i is delayed in our model, the firm can invest later in this technology θ_i against a lower cost than if it had invested at time t_0. Therefore the value of the option to postpone the adoption of technology θ_i will be positive. Consequently, the critical switching level of I determined by the net present value method will be larger than that determined by equation (3.16).

The following holds for the net present value switching level I_i^{NPV} for switching from technology ζ_0 to technology θ_i:

$$\frac{\pi(\zeta_0)}{r} = \frac{\pi(\theta_i)}{r} - I_i^{NPV}. \tag{3.31}$$

We can rewrite (3.31) as follows:

$$I_i^{NPV} = \frac{\pi(\theta_i) - \pi(\zeta_0)}{r}. \tag{3.32}$$

If we compare (3.32) with (3.16) it is not hard to see that the optimal switching level is smaller than the one obtained according to the net present value method. This implies that it is optimal to switch later, i.e. the firm anticipates at the decrease of the investment cost. As with a financial call option it is optimal to wait with exercising until the option is sufficiently deep in the money. Note that when the investment cost equals the net present value switching level, the investment option is at the money.

When the investment cost equals the critical level, the discounted gains from investing (the change in profit flow) with discounting rate $\rho (> 0)$ are exactly offset by the investment cost:

$$I_i^* = \int_{t=0}^{\infty} (\pi(\theta_i) - \pi(\zeta_0)) e^{-\rho t} dt = \frac{\pi(\theta_i) - \pi(\zeta_0)}{\rho}. \tag{3.33}$$

From (3.33) we obtain that the net present value prescribes to discount with the interest rate: $\rho = r$. According to the optimal investment strategy the discount rate must be $\rho = r + \alpha_i + \lambda\beta_i$ (cf. (3.16)). This implies that the rate at which the investment cost decreases has to be added to the interest rate.

THEOREM 3.2 *Let* $k = \arg \max\limits_{i \in \{1,\dots,N(t)\}} (F_i(t, T(t), \zeta_0))$ *and consider a time* $t\,(\geq 0)$. *Given that the firm has not invested before, according to the net present value method, the firm invests at time* t *in technology* θ_k *if the following condition is fulfilled:*

$$I_k(t) < I_k^{NPV}. \qquad (3.34)$$

Comparing this with the optimal investment strategy we see that the net present value method leads to wrong investment decisions, because it ignores (1) the decrease of the investment costs and (2) the fact that more efficient technologies will become available in the future. The first point is reflected in the critical levels (compare equation (3.16) and (3.32)). The second in the set of technologies over which is maximized (compare definitions of k in Theorems 3.1 and 3.2).

EXAMPLE 3.1 (CONTINUED) *For our parameter set it holds that*

$$I_i^{NPV} > I_0, \text{ for all } i \in \mathbf{N}.$$

This implies that according to the net present value method the firm adopts technology 1 at time T_1.

6.   CONCLUSIONS

The objective of this chapter was to analyze optimal technology adoption of a firm, while the sunk cost investments of each available technology decrease over time.

In the one switch case it is optimal for a firm to invest in a technology if three conditions are fulfilled: (1) compared to other technologies, the value of the firm is maximized by investing in that particular technology, (2) the investment cost of that technology is below its critical value (the option to invest is sufficiently deep in the money), and (3) that this particular technology is already invented.

Unfortunately, it turned out that the multiple switch case is too complex to solve.

Appendix
A.   PROOFS

PROOF OF PROPOSITION 3.1 *It holds that for* $t \in [T_i, T_{i+1})$ *the investment cost* $I_i(t)$ *is the solution of*

$$dI_i(t) = -\alpha_i I_i(t)\, dt, \qquad (3.35)$$
$$I_i(T_i) = I_{i0}. \qquad (3.36)$$

The solution of this system of equations is

$$I_i(t) = I_{i0}e^{-\alpha_i(t-T_i)}, \quad \text{for } t \in [T_i, T_{i+1}). \tag{3.37}$$

At time $t = T_{i+1}$ the investment cost instantaneously decreases with factor β_i, thus

$$
\begin{aligned}
I_i(T_{i+1}) &= I_{i0}e^{-\alpha_i(T_{i+1}-T_i)} - \beta_i I_{i0}e^{-\alpha_i(T_{i+1}-T_i)} \\
&= (1-\beta_i) I_{i0}e^{-\alpha_i(T_{i+1}-T_i)}.
\end{aligned}
\tag{3.38}
$$

Over the interval $[T_{i+1}, T_{i+2})$ the decrease of the investment cost is again given by equation (3.35), but now the initial investment cost level is given by equation (3.38). Hence, for $t \in [T_{i+1}, T_{i+2})$ we have

$$
\begin{aligned}
I_i(t) &= (1-\beta_i) I_{i0}e^{-\alpha_i(T_{i+1}-T_i)}e^{-\alpha_i(t-T_{i+1})} \\
&= (1-\beta_i) I_{i0}e^{-\alpha_i(t-T_i)}.
\end{aligned}
\tag{3.39}
$$

The investment cost at time $t = T_{i+2}$ equals

$$
\begin{aligned}
I_i(T_{i+2}) &= (1-\beta_i) I_{i0}e^{-\alpha_i(T_{i+2}-T_i)} - \beta_i(1-\beta_i) I_{i0}e^{-\alpha_i(T_{i+2}-T_i)} \\
&= (1-\beta_i)^2 I_{i0}e^{-\alpha_i(T_{i+2}-T_i)}.
\end{aligned}
\tag{3.40}
$$

The number of technology arrivals and decreases in the investment cost $I_i(t)$ on the interval $[T_i, t)$ is equal to $N(t) - i$. Repeating the steps above $N(t) - i$ times gives equation (3.3). □

PROOF OF PROPOSITION 3.2 *This proposition is just an application of Theorem 2.4 in Appendix A of Chapter 2. Note that the threshold works the other way around in this chapter. When the investment cost is above the threshold, the firm waits with investing and if the investment cost is below the threshold investing is optimal. Therefore the function mentioned in the first condition of Theorem 2.4 must be increasing in $I_i(t)$. The function is given by*

$$\pi(\zeta_0) - r\left(\frac{\pi(\theta_i)}{r} - I_i(t)\right) - \lim_{dt\downarrow 0}\frac{1}{dt}E\left[dI_i(t)|I_i(t)\right]. \tag{3.41}$$

Expanding equation (3.41) with Itô's lemma gives

$$\pi(\zeta_0) - \pi(\theta_i) + rI_i(t) + \lambda\beta_i I_i(t) + \alpha_i I_i(t), \tag{3.42}$$

which is clearly increasing in $I_i(t)$. The second condition is also fulfilled, because the investment cost is decreasing over time. Given two values of the investment cost, I_{i1} and I_{i2} such that $I_{i1} < I_{i2}$, it holds that $\Pr(I_i(t) \le I|I_{i1}) > \Pr(I_i(t) \le I|I_{i2})$. □

II

GAME THEORETIC ADOPTION MODELS

Chapter 4

ONE NEW TECHNOLOGY

1. INTRODUCTION

A feature of the last decade is that firms more and more face competition on their output markets. One reason is the abolition of monopolistic markets created by government. In the Netherlands examples are the opening of the markets for telecommunication, railway and power supply. Another reason is the, still ongoing, process of mergers, which due to legislation will not end with a market with only one supplier. The result is that markets with only one supplier and markets with many suppliers seem to disappear. Thus, in its own investment decision, a firm should take into account the investment behavior by its competitors, which is dealt with in this paper.

The existing literature on technology adoption models can be divided into two categories. The models in the first category are decision theoretic models that analyze the technology investment decision of a single firm. In the most advanced models there are multiple new technologies that arrive over time according to a stochastic process. Examples are Balcer and Lippman, 1984, Nair, 1995, Rajagopalan et al., 1998 and Farzin et al., 1998 (see also Chapters 2 and 3 of this book). These models analyze the investment decision of one firm in isolation, so that the effects of competition are not incorporated.

The second category models are game theoretic models. Two (or more) firms compete on an output market and produce goods using a particular technology. Then, a new and more efficient technology is invented, and the question is at what time the firms should adopt it. Reinganum, 1981 was the first to analyze this kind of model. She considered a duopoly with identical firms, in which there is no uncertainty in the innovation process, and one new technology is considered. The investment expenditure required to adopt the new technology decreases over time and the efficiency improvement is known. If a firm adopts the new technology before the other one does, it makes substantial profits at the expense of the other firm. On the other hand the investment cost being decreasing over time provides an incentive to wait with investing. Reinganum assumes that the firms precommit themselves to adoption times, so she automatically obtains open loop equilibria.

Fudenberg and Tirole, 1985 proved that in the open loop equilibria the leader (the firm that invests first) earns more than the follower (the firm that invests second). Since precommitment seems not to be very realistic in the strategic setting of a duopoly, Fudenberg and Tirole extended Reinganum's model by relaxing this assumption and by not determining beforehand which firm is the leader. They therefore allow firms to preempt each other. After extending the standard Nash equilibrium concept, closed loop equilibria are obtained. It turns out that the equilibria exhibit rent equalization.

The Reinganum-Fudenberg-Tirole model has been (and still is) the starting point of many technology adoption models in a duopoly setting. Hendricks, 1992 adds uncertainty to the model, by assuming that a firm is uncertain about the innovative capabilities of its rival. A firm is either an innovator or an imitator and only the firm itself knows what type it is. Hendricks assumes that an imitator can only play the follower role in the game. The result of adding this uncertainty is that in case there are large preemption gains there is no longer rent equalization in equilibrium. Firms that are innovators have an incentive to delay the adoption, since with positive probability they believe that their rival is an imitator. The advantage of adding the uncertainty is that Hendricks can apply the normal Nash equilibrium concept. Hendricks also discusses what the result is of extending the model even further by making the profitability of the technology uncertain. He argues that two cases can occur. The first is equivalent to the one described above and in the second case the firms end up in an attrition game. In an attrition game each player wants the other player to move as first. However, given that a player has to move first, the best thing for this player is to move as early as possible. We refer to Appendix A.3 for a more formal treatment of attrition games. In that appendix we use the equilibrium concepts introduced in Hendricks et al., 1988.

Hoppe, 2000 formalizes Hendricks's discussion on uncertain profitability. She starts with the Reinganum-Fudenberg-Tirole model and assumes that the innovation is either good or bad. Hoppe shows for what parameter values the model results in a preemption game or an attrition game. She does not use the equilibrium concepts introduced by Hendricks et al., 1988, that is Hoppe, does not mention the equilibria with symmetric strategies for her attrition games.

Stenbacka and Tombak, 1994 extended the Reinganum-Fudenberg-Tirole model by making the time between adoption and successful implementation stochastic. To motivate this model feature, Yorukoglu, 1998 argues that information technology capital may require significant experience to operate efficiently. For instance, an econometric study by Brynjolfsson et al., 1991 finds lags of two or three years before the organizational impacts of information technology become effective. Due to the lack of mathematical precision the paper of Stenbacka and Tombak led to two follow up papers, namely by Götz, 2000 and by Huisman and Kort, 1998a. Both papers correct mistakes of Stenbacka and Tombak. Götz studies the original model with asymmetric firms, whereas Huisman and Kort concentrate on the symmetric firm case.

The chapter is organized as follows. In Section 2 we describe and present the results of the basic technology adoption model, i.e. the model that was introduced in Reinganum, 1981 and extensively studied in Fudenberg and Tirole, 1985. The extension to the Reinganum-Fudenberg-Tirole model as presented in Stenbacka and Tombak, 1994 and studied in Götz, 2000 and in Huisman and Kort, 1998a is considered in Section 3. The last section concludes.

2. REINGANUM-FUDENBERG-TIROLE MODEL

In this section we present and analyze the game theoretic technology adoption model that was introduced by Reinganum, 1981. There are two identical firms active on an output market. The firms are labelled 1 and 2. An infinite planning horizon is considered, on which the risk-neutral firms maximize their value at discount rate r (> 0). Initially the firms produce with a technology of which the efficiency is denoted by θ_0. At time $t = 0$ a new technology, with efficiency θ_1 ($> \theta_0$) becomes available and the firms must decide when to adopt that technology. We assume that the firms do not have any market power on the technology market, that is they are one of many firms on the technology market. When a firm adopts the new technology it incurs an investment cost, I, which is a convex decreasing function of time t (≥ 0):

$$I(t) > 0, \quad \frac{\partial I(t)}{\partial t} \leq 0 \quad \text{and} \quad \frac{\partial^2 I(t)}{\partial t^2} \geq 0. \qquad (4.1)$$

There can be three reasons for the decrease of the investment cost of a particular technology: (1) it becomes old-fashioned, (2) the firms that are most eager to buy the technology have already bought it so that technology suppliers have to drop their price in order to find additional buyers, and (3) due to learning by doing the technology supplier can produce the technology in a cheaper way.

The profit function of a firm i is denoted by $\pi\left(\theta_i, \theta_j\right)$ where θ_i and θ_j are the efficiencies of the technologies in use by firm i and firm j, respectively, where $i, j \in \{1, 2\}$ and $i \neq j$.

There is a first mover advantage in the sense that the gains for being first to adopt the new technology are higher than for being second:

$$\pi\left(\theta_1, \theta_0\right) - \pi\left(\theta_0, \theta_0\right) > \pi\left(\theta_1, \theta_1\right) - \pi\left(\theta_0, \theta_1\right) \geq 0, \tag{4.2}$$

The four possible profit flows are assumed to be ranked in the following way:

$$\pi\left(\theta_1, \theta_0\right) > \pi\left(\theta_1, \theta_1\right) \geq \pi\left(\theta_0, \theta_0\right) \geq \pi\left(\theta_0, \theta_1\right) \geq 0. \tag{4.3}$$

Thus, we assume that a firm can make higher profits when it uses a more efficient technology itself and when its rival uses a less efficient technology. The following assumption rules out immediate adoption:

$$rI\left(0\right) - \left.\frac{\partial I\left(t\right)}{\partial t}\right|_{t=0} > \pi\left(\theta_1, \theta_0\right) - \pi\left(\theta_0, \theta_0\right). \tag{4.4}$$

Equation (4.4) states that, at time $t = 0$, the marginal costs of adoption are larger than the marginal benefits of adopting. The costs of adoption at time t are equal to

$$-I\left(t\right) e^{-rt}, \tag{4.5}$$

so that the marginal costs equal

$$\left(rI\left(t\right) - \frac{\partial I\left(t\right)}{\partial t}\right) e^{-rt}. \tag{4.6}$$

The firm that invests first is called the leader and the other firm is called the follower.

In Subsection 2.1 the open loop equilibrium of the model is presented. The feedback equilibrium is discussed in Subsection 2.2. In these first two subsections the firm roles (leader or follower) are assigned exogenous to the firms. This assumption is relaxed in Subsection 2.3.

2.1   OPEN LOOP EQUILIBRIUM

In an open loop equilibrium both firms precommit themselves to an adoption time at the beginning of the game, i.e. the firms do not take into account that they can influence the other firm's adoption time. The adoption time of the leader and follower are denoted by t_L and t_F, respectively. By definition, the leader adopts at the same time or before the follower, i.e. $0 \leq t_L \leq t_F$. Given the adoption times t_L and t_F, the value of the leader equals

$$
\begin{aligned}
V_L\left(t_L, t_F\right) &= \int_{t=0}^{t_L} \pi\left(\theta_0, \theta_0\right) e^{-rt} dt + \int_{t=t_L}^{t_F} \pi\left(\theta_1, \theta_0\right) e^{-rt} dt \\
&\quad + \int_{t=t_F}^{\infty} \pi\left(\theta_1, \theta_1\right) e^{-rt} dt - I\left(t_L\right) e^{-rt_L} \\
&= \frac{\pi\left(\theta_0, \theta_0\right)}{r} + e^{-rt_L}\left(\frac{\pi\left(\theta_1, \theta_0\right) - \pi\left(\theta_0, \theta_0\right)}{r} - I\left(t_L\right)\right) \\
&\quad + e^{-rt_F}\left(\frac{\pi\left(\theta_1, \theta_1\right) - \pi\left(\theta_1, \theta_0\right)}{r}\right).
\end{aligned}
\tag{4.7}
$$

In the same way the value of the follower, $V_F\left(t_L, t_F\right)$, can be derived. The follower's value is given by

$$
\begin{aligned}
V_F\left(t_L, t_F\right) &= \frac{\pi\left(\theta_0, \theta_0\right)}{r} + e^{-rt_L}\left(\frac{\pi\left(\theta_0, \theta_1\right) - \pi\left(\theta_0, \theta_0\right)}{r}\right) \\
&\quad + e^{-rt_F}\left(\frac{\pi\left(\theta_1, \theta_1\right) - \pi\left(\theta_0, \theta_1\right)}{r} - I\left(t_F\right)\right).
\end{aligned}
\tag{4.8}
$$

The open loop equilibria are equal to the intersection points of the reaction curves of the follower and the leader. The reaction function for the follower is derived by calculating, for each fixed adoption time t_L of the leader, the adoption time of the follower, $R_F\left(t_L\right)$, that maximizes the follower's value, taking into account the fact that by definition the follower has to adopt after the leader: $R_F\left(t_L\right) \geq t_L$. In the same way the reaction function of the leader is derived by taking a fixed adoption time t_F of the follower and deriving the best reply of the leader $R_L\left(t_F\right)$, under the condition that the leader has to adopt before the follower: $R_L\left(t_F\right) \leq t_F$.

The procedure described above implies that the reaction curves are built up by the first order conditions of maximizing the value functions and the 45 degree line. The 45 degree line is the line that resembles joint-adoption of the leader and the follower. Note that we do not claim that the reaction functions are continuous. In fact, in Section 3 we present an example where the reaction functions are not continuous.

The first order condition for maximizing $V_F(t_L, t_F)$ over t_F is

$$\pi(\theta_1, \theta_1) - \pi(\theta_0, \theta_1) - rI(t_F) + \left.\frac{\partial I(t)}{\partial t}\right|_{t=t_F} = 0. \qquad (4.9)$$

Let T_F be the solution of equation (4.9). Equations (4.1), (4.2), and (4.4) ensure the existence of a unique positive maximum. The reaction function of the follower is given by

$$R_F(t_L) = \begin{cases} T_F & \text{if } t_L \le T_F, \\ t_L & \text{if } t_L > T_F. \end{cases} \qquad (4.10)$$

The value $V_L(t_L, t_F)$ of the leader is maximized with respect to t_L if

$$\pi(\theta_1, \theta_0) - \pi(\theta_0, \theta_0) - rI(t_L) + \left.\frac{\partial I(t)}{\partial t}\right|_{t=t_L} = 0. \qquad (4.11)$$

The solution of equation (4.11) is denoted by T_L. As with T_F, equations (4.1), (4.2) and (4.4) ensure the existence of a unique positive maximum. Further, these equations imply that $0 < T_L < T_F$. Therefore, we conclude that there is diffusion in the open loop equilibrium timings. The leader's reaction function equals

$$R_L(t_F) = \begin{cases} t_F & \text{if } t_F < T_L, \\ T_L & \text{if } t_F \ge T_L. \end{cases} \qquad (4.12)$$

In Figure 4.1 the reaction curves of the leader and follower are plotted. The analysis above results in the following proposition that summarizes

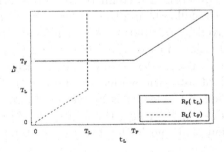

Figure 4.1.   Reaction curves of leader and follower.

the open loop equilibrium. For a more formal proof we refer to Reinganum, 1981.

PROPOSITION 4.1 *The open loop equilibrium with exogenous firm roles is as follows. The leader adopts the technology at time T_L (> 0) and the follower adopts the technology at time T_F ($> T_L$), where T_L and T_F are found by solving equations (4.11) and (4.9), respectively.*

2.2 FEEDBACK EQUILIBRIUM

In a feedback equilibrium the leader takes into account that its investment decision affects the decision of the follower. The follower's reaction is the same as in the open loop case. To determine the equilibrium we plot the leader's payoff as function of its own adoption date and take the adoption date of the follower equal to its optimal reaction. The equilibrium is given by the adoption time at which the leader's payoff is at its maximum and the optimal reaction of the follower on that adoption date.

Define the following three functions

$$L(t) = V_L(t, R_F(t)), \tag{4.13}$$
$$F(t) = V_F(t, R_F(t)), \tag{4.14}$$
$$M(t) = V_L(t, t). \tag{4.15}$$

The function $L(t)$ ($F(t)$) is equal to the expected discounted value at time $t = 0$ of the leader (follower) when the leader invests at time t. $M(t)$ resembles the discounted value at time $t = 0$ of the firm when there is joint-adoption at time t.

The definition of T_F implies that joint-adoption is not optimal before T_F, i.e.

$$M(t) < L(t), \quad t < T_F, \tag{4.16}$$
$$M(t) < F(t), \quad t < T_F, \tag{4.17}$$

and is optimal after time T_F:

$$L(t) = F(t) = M(t), \quad t \geq T_F. \tag{4.18}$$

Define T_C to be equal to

$$T_C = \arg\max_{t \geq 0} M(t). \tag{4.19}$$

Thus T_C is the solution of

$$\pi(\theta_1, \theta_1) - \pi(\theta_0, \theta_0) - rI(T_C) + \left.\frac{\partial I(t)}{\partial t}\right|_{t=T_C} = 0. \tag{4.20}$$

Equations (4.1)-(4.3) guarantee that T_C is positive, exists, and maximizes M. These equations imply that $T_C > T_F$.

From the definitions of L, F, and M we derive that F and M are increasing on the interval $[0, T_F]$ and that L is increasing on the interval $[0, T_L)$ and decreasing on the interval $(T_L, T_F]$. Further, M is increasing on the interval (T_F, T_C) and decreasing for $t > T_C$. Hence, there exist three cases. In case A it holds that $L(T_L) > M(T_C)$. Case B is characterized by $L(T_L) < M(T_C)$ and case C by $L(T_L) = M(T_C)$. In Figures 4.2 and 4.3 the three functions are plotted for cases A and B, respectively.

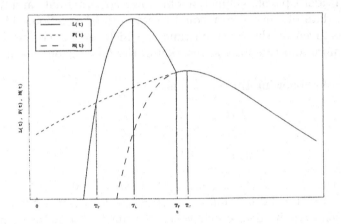

Figure 4.2.   Case A: $L(T_L) > M(T_C)$.

The analysis above implies that there are two candidates for the feedback equilibrium: (1) (T_L, T_F) and (2) (T_C, T_C). The following proposition summarizes the analysis.

PROPOSITION 4.2 *If $L(T_L) > M(T_C)$ the equilibrium is as follows: the leader adopts the technology at time T_L and the follower adopts the technology at time T_F. If $L(T_L) < M(T_C)$ the equilibrium is of the joint-adoption type, where the leader and the follower adopt the technology at time T_C. If $L(T_L) = M(T_C)$ both equilibria exist.*

2.3   ENDOGENOUS FIRM ROLES

In this subsection the firm roles will no longer be exogenously given. This is more realistic since generally firm roles will not be assigned beforehand in practice. This implies that a firm can (only) become leader

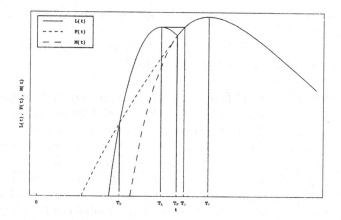

Figure 4.3. Case B: $L\left(T_L\right) < M\left(T_C\right)$.

by investing as first. Fudenberg and Tirole, 1985 were the first to analyze the model of Reinganum with endogenous firm roles. The reason
of doing this further analysis is that the payoff to the leader exceeds
the payoff to the follower in case A, i.e. $L\left(T_L\right) > F\left(T_L\right)$. Therefore it
is in each firm's interest to be the leader and end up with the higher
payoff. In order to incorporate this feature of endogenous firm roles in
the equilibrium concept, Fudenberg and Tirole developed the so-called
perfect equilibrium concept for timing games (see also Appendix A).

We analyze this problem by using Figures 4.2 and 4.3. We add case
C to case B. Thus, case B is, from now on, characterized by $L\left(T_L\right) \leq$
$M\left(T_C\right)$. Consider case A. Both firms want to become leader and adopt
at time T_L. As a result a firm will try to preempt the other firm by
investing at time $T_L - \varepsilon$, but then the other will try to preempt by
adopting at time $T_L - 2\varepsilon$ and so forth and so on. This process stops at
time T_P, where time T_P is defined as:

$$T_P = \min\left(t \in (0, T_L) \,|\, L\left(t\right) = F\left(t\right)\right). \qquad (4.21)$$

Thus, the preemption process stops at the time at which the expected
values of the leader and follower are equal. This phenomenon is called
rent equalization. In the equilibrium of case A, one of the firms preempts
at time T_P and the other firm will react by adopting at time $T_F > T_P$.

To analyze case B, first define time T_S:

$$T_S = \min\left(t \in (T_F, T_C] \,|\, M\left(t\right) = L\left(T_L\right)\right). \qquad (4.22)$$

In case B there are multiple equilibria, which can be split up in two
classes. The first class consists of the (T_P, T_F) diffusion equilibria. The

second class is a continuum of joint-adoption outcomes indexed by the date of adoption $t \in [T_S, T_C]$.

We summarize the analysis in the following proposition (see also Proposition 2 of Fudenberg and Tirole, 1985).

PROPOSITION 4.3 *(A) If $L(T_L) > M(T_C)$ there exists a unique equilibrium distribution over outcomes. With probability one-half, firm 1 adopts at time T_P and firm 2 adopts at T_F, and with probability one-half the roles of the firms are reversed. Thus the equilibrium exhibits diffusion; and with probability one the adoption dates are T_P and T_F.*
(B) If $L(T_L) \leq M(T_C)$ two classes of equilibria exist. The first class are the (T_P, T_F) diffusion equilibria. The second class is a continuum of joint-adoption outcomes indexed by the date of adoption $t \in [T_S, T_C]$.

The proof can be found in Fudenberg and Tirole, 1985. The equilibrium strategies that result in the mentioned equilibria are given in Appendix A. Fudenberg and Tirole prove that the probability of a mistake in the diffusion equilibrium, i.e. that both firms adopt simultaneously at time T_P which leads to very low profits, is zero. They derive the following properties of the equilibria.

PROPOSITION 4.4 *Joint-adoption equilibria are Pareto-ranked by their date of adoption, later adoption being more efficient from the firm's point of view.*

The implication of Proposition 4.4 is that in case B the most reasonable outcome to expect is the joint-adoption at time T_C, because it Pareto-dominates all other equilibria.

3.   STENBACKA AND TOMBAK'S EXTENSION

In this section we analyze the extension to the Reinganum-Fudenberg-Tirole model introduced in Stenbacka and Tombak, 1994. This extension is also studied in Götz, 2000 and Huisman and Kort, 1998a. The time between adoption and successful implementation is uncertain and assumed to be exponentially distributed with rate λ. Note that, contrary to Stenbacka and Tombak, we assume that the hazard rates of the firms are the same.

3.1   OPEN LOOP EQUILIBRIUM

The expected value of the leader if the leader adopts at time $t_L (\geq 0)$ and the follower adopts at time $t_F (\geq t_L)$ equals (see Stenbacka and

Tombak, 1994)

$$
\begin{aligned}
E\left[V_L\left(t_L, t_F\right)\right] = \ & \frac{\pi(\theta_0, \theta_0)}{r} + \frac{\lambda(\pi(\theta_1, \theta_0) - \pi(\theta_0, \theta_0))}{r(r+\lambda)} e^{-rt_L} \\
& + \frac{\lambda(\pi(\theta_1, \theta_1) - \pi(\theta_1, \theta_0))}{r(r+\lambda)} e^{-rt_F} \\
& + \frac{\lambda(\pi(\theta_0, \theta_1) - \pi(\theta_1, \theta_1) - \pi(\theta_0, \theta_0) + \pi(\theta_1, \theta_0))}{(r+\lambda)(r+2\lambda)} e^{-(r+\lambda)t_F + \lambda t_L} \\
& - I\left(t_L\right) e^{-rt_L}.
\end{aligned}
\tag{4.23}
$$

The expected value of the follower is given by

$$
\begin{aligned}
E\left[V_F\left(t_L, t_F\right)\right] = \ & \frac{\pi(\theta_0, \theta_0)}{r} + \frac{\lambda(\pi(\theta_0, \theta_1) - \pi(\theta_0, \theta_0))}{r(r+\lambda)} e^{-rt_L} \\
& + \frac{\lambda(\pi(\theta_1, \theta_1) - \pi(\theta_0, \theta_1))}{r(r+\lambda)} e^{-rt_F} \\
& + \frac{\lambda(\pi(\theta_0, \theta_1) - \pi(\theta_1, \theta_1) - \pi(\theta_0, \theta_0) + \pi(\theta_1, \theta_0))}{(r+\lambda)(r+2\lambda)} e^{-(r+\lambda)t_F + \lambda t_L} \\
& - I\left(t_F\right) e^{-rt_F}.
\end{aligned}
\tag{4.24}
$$

Differentiating equation (4.23) with respect to t_L gives

$$
\begin{aligned}
& \frac{\lambda(\pi(\theta_1, \theta_0) - \pi(\theta_0, \theta_0))}{r+\lambda} - rI\left(t_L\right) + \left.\frac{\partial I(t)}{\partial t}\right|_{t=t_L} \\
& - \frac{\lambda^2(\pi(\theta_0, \theta_1) - \pi(\theta_1, \theta_1) - \pi(\theta_0, \theta_0) + \pi(\theta_1, \theta_0))}{(r+\lambda)(r+2\lambda)} e^{-(r+\lambda)(t_F - t_L)} = 0.
\end{aligned}
\tag{4.25}
$$

When, given t_F, there exists a solution of equation (4.25) such that it is smaller or equal to t_F, this solution is denoted by $\tau_L\left(t_F\right)$. Define $\widehat{\tau}_L = \sup\left(t \geq 0 \mid \tau_L\left(t\right) = t\right)$. Then

$$
R_L\left(t_F\right) = \begin{cases} t_F & \text{if } t_F < \widehat{\tau}_L, \\ \tau_L\left(t_F\right) & \text{if } t_F \geq \widehat{\tau}_L. \end{cases}
\tag{4.26}
$$

Differentiating equation (4.24) with respect to t_F gives the following first order condition

$$
\begin{aligned}
& \frac{\lambda(\pi(\theta_1, \theta_1) - \pi(\theta_0, \theta_1))}{r+\lambda} - rI\left(t_F\right) + \left.\frac{\partial I(t)}{\partial t}\right|_{t=t_F} \\
& + \frac{\lambda(\pi(\theta_0, \theta_1) - \pi(\theta_1, \theta_1) - \pi(\theta_0, \theta_0) + \pi(\theta_1, \theta_0))}{(r+2\lambda)} e^{-\lambda(t_F - t_L)} = 0.
\end{aligned}
\tag{4.27}
$$

The solution of the last equation (if it exists) is denoted by $\tau_F\left(t_L\right)$. We define $T_F = \inf\left(t \geq 0 \mid \tau_F\left(t\right) = t\right)$, i.e. T_F is the first point in time for which the optimal reaction for the follower to adoption of the leader at time t_L is also adopting at time t_L. It is obvious that the reaction function of the follower is equal to the 45 degree line for all $t_L > T_F$. This results in the following expression for the reaction curve of the follower

$$
R_F\left(t_L\right) = \begin{cases} \tau_F\left(t_L\right) & \text{if } t_L \leq T_F, \\ t_L & \text{if } t_L > T_F. \end{cases}
\tag{4.28}
$$

Götz, 2000 shows that there always exists a unique date t_S^* which satisfies

$$t_S^* = \tau_F\left(t_S^*\right) = \tau_L\left(t_S^*\right). \qquad (4.29)$$

Götz also proves that $\left(t_S^*, t_S^*\right)$ is not an open loop equilibrium for λ larger than a certain threshold λ_0. That is, the point $\left(t_S^*, t_S^*\right)$ is not contained in the set of intersection points of the reaction curves. The reason is that, for at least one player, the point $\left(t_S^*, t_S^*\right)$ is not on its reaction curve, because the second order condition is not satisfied for that point. Assume that the point $\left(t_S^*, t_S^*\right)$ is not on the reaction curve of the follower. Then the first order condition of the follower, given that the leader adopts at time t_S^*, has two solutions: t_S^* and $R_F\left(t_S^*\right)$. The second order condition is satisfied for the point $\left(t_S^*, R_F\left(t_S^*\right)\right)$. This implies that the reaction curve of the follower is discontinuous at time T_F,

$$\lim_{\xi \downarrow 0} R_F\left(T_F - \xi\right) = \tau_F\left(T_F\right) > T_F, \qquad (4.30)$$

while

$$E\left[V_F\left(T_F, T_F\right)\right] = E\left[V_F\left(T_F, \tau_F\left(T_F\right)\right)\right]. \qquad (4.31)$$

Thus, if the leader adopts at time T_F, the follower is indifferent between adopting at time $\tau_F\left(T_F\right)$ and adopting at time T_F. Suppose that the follower chooses T_F, then it is in the leader's interest to adopt at time $\tau_L\left(T_F\right)\left(< T_F\right)$, so that $\left(T_F, T_F\right)$ is not an equilibrium. Therefore, the equilibrium is of the diffusion type in this case. Note that this is consistent with Reinganum, 1981 ($\lambda = \infty$). The timing of the diffusion equilibrium in this case, $\left(t_L^*, t_F^*\right)$, is the solution of

$$\begin{cases} R_F\left(t_L\right) = t_F, \\ R_L\left(t_F\right) = t_L. \end{cases} \qquad (4.32)$$

For $\lambda \leq \lambda_0$ the reaction functions are continuous and $R_F\left(t_S^*\right) = t_S^* = T_F$, thus $\left(T_F, T_F\right)$ is a solution of (4.32). Here, there is certainly a joint-adoption equilibrium and there can also be a diffusion equilibrium.

The result above implies the incorrectness of part (a) of Proposition 1 of Stenbacka and Tombak, 1994 on p. 399: *In an open loop equilibrium the extent of dispersion between the adoption timings will be increased if the degree of uncertainty is increased.* This part of the proposition is incorrect if we find a parameter setting for which the equilibrium is of the diffusion type for large λ and of the joint-adoption type for small λ. The following example contradicts Stenbacka and Tombak's Proposition 1 in this way, which is also mentioned in Götz, 2000. In the same example it is also shown that part (b) of Stenbacka and Tombak's Proposition 1 is incorrect: *In an open loop equilibrium the extent of dispersion between the adoption timings will be increased if the advantages of being the first to succeed decrease relative to the gains from being the second to succeed.* This is not mentioned in Götz, 2000.

EXAMPLE 4.1 *We complete the example Stenbacka and Tombak use to illustrate case (b) of their Proposition 1 and which Götz has extended. Stenbacka and Tombak start out with a Cournot duopoly model with linear inverse demand function $p = a - q_1 - q_2$, constant marginal costs, c, an innovation that reduces the marginal cost to $c - \varepsilon$. It can be derived that*

$$\pi(\theta_1, \theta_0) = \frac{1}{9}(a - c + 2\varepsilon)^2,$$

$$\pi(\theta_1, \theta_1) = \frac{1}{9}(a - c + \varepsilon)^2,$$

$$\pi(\theta_0, \theta_0) = \frac{1}{9}(a - c)^2,$$

$$\pi(\theta_0, \theta_1) = \frac{1}{9}(a - c - \varepsilon)^2.$$

As in Götz, 2000, we take the investment cost I at time t equal to I(t) $= 1000e^{-0.2t}$, $a = 9$, $\varepsilon = 1$, $c = 1$ and the discount rate $r = 0.05$. In Figure 4.4 the reaction curves of the leader and the follower are plotted for $\lambda = 1$ and $\lambda = 3$. Figure 4.5 shows the reaction curves of the leader and the follower for $\lambda = 4$. Note that Götz also plots reaction curves, but for the individual firms. Where we assign the leader and follower role beforehand, Götz has to do it afterwards. He only plots the reaction functions for $\lambda = 1$ and $\lambda = 3$. He argues that there is not a joint-adoption equilibrium for $\lambda = 4$, but he does not plot the reaction curves for this case. Note that for $\lambda = 4$ the reaction curves are discontinuous. Thus in this example the threshold λ_0 will be somewhere within the interval (3, 4). In Table 4.1 the equilibria for the three different scenarios are summarized. Comparing the third and the first plot we see that increasing the uncertainty (lower λ) leads to a decreased extent of dispersion of the adoption timings, which implies the incorrectness of part (a) of Proposition 1 of Stenbacka and Tombak, 1994.

In the next scenario we take $\lambda = 4$ and $\varepsilon = 0.5$. The reaction functions are plotted in the right panel of Figure 4.5 and in Table 4.2 the equilibria are listed. By comparing the fourth and third scenario we see that an increase of ε (the cost reduction from successful implementation) results

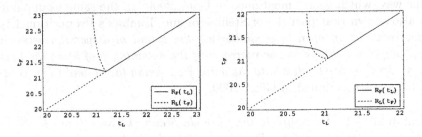

Figure 4.4. Reaction curves for $\lambda = 1$ (left panel) and $\lambda = 3$ (right panel).

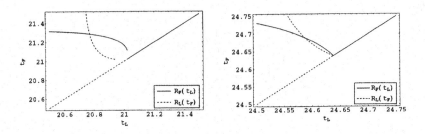

Figure 4.5. Reaction curves for $\lambda = 4$ and for $\varepsilon = 1$ (left panel) and $\varepsilon = 0.5$ (right panel).

Table 4.1. Equilibria for different values of λ and $\varepsilon = 1$.

λ	(t_L^*, t_F^*)	(t_S^*, t_S^*)
1	none	$(21.20, 21.20)$
3	$(20.83, 21.26)$	$(21.04, 21.04)$
4	$(20.77, 21.29)$	none

in an increase of the extent of dispersion, which disproves part (b) of Stenbacka and Tombak's Proposition 1.

3.2   FEEDBACK EQUILIBRIUM

Like in Subsection 2.2, to determine the feedback equilibrium we plot the leader's payoff as function of its own adoption date and take the adoption date of the follower equal to its optimal reaction.

Table 4.2. Equilibria for $\lambda = 4$ and $\varepsilon = 0.5$.

(t_L^*, t_F^*)	(t_S^*, t_S^*)
$(24.59, 24.69)$	$(24.64, 24.64)$

Another method to find the feedback equilibrium is used by Götz, 2000. He also starts with the reaction function of the follower and plots isoprofit curves of the leader in the same figure. The equilibrium is found by the point that is on the follower's reaction curve and yields the highest profit for the leader.

The method to derive the feedback equilibrium, used by Stenbacka and Tombak, 1994 will not lead to the right solutions. As in the open loop case they draw reaction curves for both the leader and follower and consider the intersections as equilibria. However, in the feedback case the leader affects the follower's decision, which implies that the leader does not take a follower's decision for granted. This makes it impossible to draw a reaction curve for the leader. This explains the different results between Stenbacka and Tombak on the one side and Götz and this chapter on the other side. Stenbacka and Tombak (will) only find dispersed leader-follower adoption times, whereas we will show that joint-adoption can also be a feedback equilibrium. If we find an equilibrium with dispersed timings, these timings are a solution of the system of first order conditions of Stenbacka and Tombak. See Appendix B for a formal proof of the fact that the system of first order conditions does not have a joint-adoption solution. This contradicts the claim by Stenbacka and Tombak, 1994 that there also exists a joint-adoption solution for that system of equations. The reasons are: (1) the only feasible pairs of timings are the ones on the reaction curve of the follower, the timings being dispersed, which implies that the reaction curve of the follower equals the first order condition of the follower, and (2) the value of the leader must be maximized, which gives the first order condition of the leader.

Define the following three functions

$$L(t) = E[V_L(t, R_F(t))], \tag{4.33}$$
$$F(t) = E[V_F(t, R_F(t))], \tag{4.34}$$
$$M(t) = E[V_L(t, t)]. \tag{4.35}$$

The function $L(t)$ $(F(t))$ is equal to the expected discounted value at time zero of the leader (follower) when the leader invests at time t. $M(t)$

resembles the value of the firm when there is joint-adoption at time t. Remember that the best reply adoption time of the follower when the leader invests at time t is denoted by $R_F(t)$.

Joint-adoption is by definition not optimal before T_F, i.e.

$$M(t) < L(t), \quad t < T_F, \tag{4.36}$$

$$M(t) < F(t), \quad t < T_F, \tag{4.37}$$

and is optimal after time T_F:

$$L(t) = F(t) = M(t), \quad t \geq T_F. \tag{4.38}$$

As long as $R_F(t) > t$, the expected payoff of the follower increases when the leader adopts later. This is because a later adoption date for the leader implies that the expected implementation date of the leader will also be later, which makes him a less strong competitor, so that the follower can reach higher profits. Hence, on the interval $[0, T_F)$ the function F is increasing in t. On that interval, the leader curve is first increasing and then decreasing in t. This is a result of the following proposition (see also Proposition 2 of Stenbacka and Tombak, 1994).

PROPOSITION 4.5 *It holds that*

$$T_L < t_L^* \leq t_F^* < R_F(T_L), \tag{4.39}$$

where $T_L = \arg \max_{t \in [0, T_F)} L(t)$ *and* (t_L^*, t_F^*) *is the open loop equilibrium.*

Note that these T_L and $R_F(T_L)$ are the solutions of the system of first order conditions of Stenbacka and Tombak, 1994 in the feedback case. Intuitively the proposition is understandable since the reaction function of the follower is such that $\frac{\partial t_F}{\partial t_L} < 0$ (differentiate equation (4.27) with respect to t_L). Hence, the leader knows that when it adopts earlier the follower will adopt later. This strategic interaction, which is present in the feedback setup, thus leads to a larger expected time interval on which the leader collects first implementor profits. Compared to open loop, this gives an extra incentive for the leader to adopt earlier. This tendency to adopt earlier, however, is tempered by the increase in investment costs. Since the proof of Stenbacka and Tombak, 1994 is not correct, we give a correct proof in Appendix B. The reasons why their proof is not correct are: (1) the so-called additional terms in Stenbacka and Tombak's equation (11) (cf. p. 401 of Stenbacka and Tombak, 1994) are not all negative, and (2) these three terms are dependent on T_L. This implies that a more careful mathematical treatment is needed to prove the claim.

If the reaction function of the follower is discontinuous at time T_F (e.g. see the left panel of Figure 4.5) the leader curve makes a discontinuous jump downwards at time T_F. The reason is that before T_F the follower adopts later than the leader so that the leader enjoys higher profits than the follower during a time interval with positive length in expectation, while from T_F onwards expected leader and follower value are equal.

Stenbacka and Tombak, 1994 define T_C to be equal to

$$T_C = \arg\max_{t \geq 0} M(t). \tag{4.40}$$

They call T_C the cooperative adoption time, but we will argue below that this can also be the equilibrium adoption time of the non-cooperative game. Since for all $t \in [0, T_F]$: (1) $F(t) > M(t)$, (2) $F(t)$ is increasing in t, and (3) $F(T_F) = M(T_F)$, it holds that $T_C \geq T_F$. Differentiating $E[V_L(t, t)]$ with respect to t gives the following equation that defines T_C implicitly:

$$\frac{\lambda(2\lambda\pi(\theta_1, \theta_1) + r(\pi(\theta_1, \theta_0) + \pi(\theta_0, \theta_1)) - 2(r + \lambda)\pi(\theta_0, \theta_0))}{(r + \lambda)(r + 2\lambda)}$$

$$-rI(T_C) + \left.\frac{\partial I(t)}{\partial t}\right|_{t = T_C} = 0. \tag{4.41}$$

The second order condition for T_C to be a maximum is given by

$$r \left.\frac{\partial I(t)}{\partial t}\right|_{t = T_C} - \left.\frac{\partial^2 I(t)}{\partial t^2}\right|_{t = T_C} < 0, \tag{4.42}$$

which is satisfied due to equation (4.1).

The analysis above implies that there are two candidates for the feedback equilibrium: (1) $(T_L, R_F(T_L))$ and (2) (T_C, T_C). The following proposition summarizes the analysis.

PROPOSITION 4.6 *There are three different equilibrium scenarios.*

- *If $E[V_L(T_L, R_F(T_L))] > E[V_L(T_C, T_C)]$ the equilibrium is as follows: the leader adopts the technology at time T_L and the follower adopts the technology at time $R_F(T_L)$.*

- *If $E[V_L(T_L, R_F(T_L))] < E[V_L(T_C, T_C)]$ the equilibrium is of the joint adoption type, where the leader and the follower adopt the technology at time T_C.*

- *If $E[V_L(T_L, R_F(T_L))] = E[V_L(T_C, T_C)]$ both of the two described equilibria exist.*

EXAMPLE 4.1 (CONTINUED) *In Figures 4.6-4.7 we have plotted the pay-off functions, L (t), F (t), and M (t) for the first three scenarios of the example of the previous section. In Table 4.3 we have summarized the equilibrium timings. Note that the feedback equilibrium (T_C, T_C) is not an open loop equilibrium. The reason is that if the leader knows for sure that the follower is going to adopt at time T_C he can do better by adopting before T_C. See also Figure 4.8 in which we have plotted the value of the leader as function of its own adoption time under the assumption that the follower is going to adopt at time T_C for sure $(\lambda = 1)$. We may conclude that adopting at time T_C (= 24.64) is not optimal for the leader. But still the strategic interaction makes (T_C, T_C) the feedback equilibrium. This shows the incorrectness of the construction of the equilibria of Stenbacka and Tombak: (T_C, T_C) will not be on their "leader's reaction curve", whereas it is the equilibrium. From the previous section we know that for $\lambda = 4$ the reaction curve of the follower is discontinuous at time T_F, in the right panel of Figure 4.7 we have zoomed at that point to show that the leader payoff curve is indeed discontinuous at that point in time.*

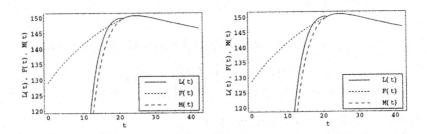

Figure 4.6.   Payoff curves for $\lambda = 1$ (left panel) and $\lambda = 3$ (right panel).

Table 4.3.   Equilibria for different values of λ.

λ	(T_C, T_C)	$M(T_C)$
1	$(24.64, 24.64)$	150.7
3	$(24.50, 24.50)$	151.0
4	$(24.48, 24.48)$	151.0

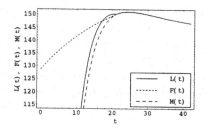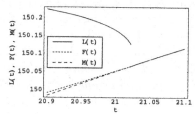

Figure 4.7. Payoff curves for $\lambda = 4$.

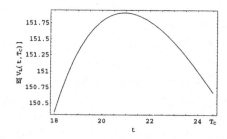

Figure 4.8. Expected value of the leader given that the follower adopts at time T_C.

3.3 ENDOGENOUS FIRM ROLES

In this section we make the firm roles endogenous. This means that both firms have equal chances to become the leader. Of course, the firm that invests first actually becomes the leader. The analysis of Subsection 2.3 can also be applied to the payoff functions defined by equations (4.33)-(4.35). Since there are no changes we refer to that section for the formal statement and the properties of the equilibria.

EXAMPLE 4.1 (CONTINUED) *In all of the three parameter settings of the example of Section 4 we are in case B. In Table 4.4 we have summarized the characteristics of the equilibria. If we change the parameters* $\pi(\theta_1, \theta_0)$ *and r into* $\pi(\theta_1, \theta_0) = 15$ *and* $r = 0.08$ *we are in case A. The equilibrium timings and payoffs are given in Table 4.5 and the leader, follower and joint-adoption curves are plotted in Figure 4.9. Note that* the equilibrium payoffs $L(T_P)$ are lower than $M(\infty) = \frac{\pi(\theta_0,\theta_0)}{r} = 88.89$, which is the payoff to each firm if both keep on producing with the old technology. This striking result means that both firms can do better by sticking to producing with their old technology forever, provided that their competitor does the same. Still, strategic interactions drive them to the preemption equilibrium just mentioned, so that these interactions have a disastrous effect on both firms' performance.

Table 4.4.   Characteristics of equilibria for different values of λ.

λ	$(T_P, R_F(T_P))$	$L(T_P)$	$(T_L, R_F(T_L))$	T_S	$M(T_S)$	T_C	$M(T_C)$
1	$(17.29, 21.50)$	147.3	$(20.63, 21.36)$	21.32	150.0	24.64	150.7
3	$(17.13, 21.35)$	147.5	$(20.54, 21.32)$	21.34	150.2	24.50	151.0
4	$(17.11, 21.33)$	147.6	$(20.58, 21.31)$	21.35	150.3	24.48	151.0

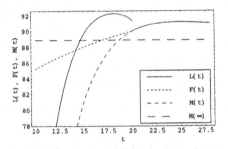

Figure 4.9.   Payoff curves for $\pi(\theta_1, \theta_0) = 15$, $\lambda = 1$, and $r = 0.08$.

Table 4.5.   Equilibrium for $\pi(\theta_1, \theta_0) = 15$, $\lambda = 1$, and $r = 0.08$.

$(T_P, R_F(T_P))$	$L(T_P)$	(T_L, T_F)	$L(T_L)$	T_C	$M(T_C)$
$(14.18, 22.22)$	87.66	$(18.10, 22.16)$	92.27	24.96	91.20

4.   CONCLUSIONS

We conclude that the introduction of uncertainty does not change the main results derived by Fudenberg and Tirole, 1985. In the feedback framework with endogenous firm roles we end up with the same two cases.

A point for further research is to extend the analysis to a model with unequal hazard rates. This would lead us back to the original and very interesting setup of Stenbacka and Tombak, 1994. The necessity for the review of their analysis and results should be clear by now.

Appendices

A.   TIMING GAMES WITH TWO IDENTICAL PLAYERS

We restrict our attention as much as possible to equilibria with symmetric strategies. The reason is that in a game with identical players, identical strategies are the most logical ones. Timing games can be divided in two classes: preemption games and attrition games.

In a timing game there are two players, 1 and 2, that have to decide when to make a single move at some time t in the interval $[0, 1]$. This is without loss of generality. For example, a game with an infinite horizon can be transformed into this framework by a change of the time variable, take $t = \frac{u}{u+1}$, where $u \in [0, \infty)$. If we denote one player by i then the other player is denoted by j.

The player that moves first is called the leader and its payoff equals $L(t)$, and the other player is called the follower and earns $F(t)$. If both players move simultaneously at time t they both get a payoff equal to $M(t)$.

We assume that the payoff functions $L(t)$ and $F(t)$ are continuous on the time interval $[0, 1]$. When one (or both) of the payoff functions is discontinuous (as in Example 4.1 for $\lambda = 4$ and $\varepsilon = 1$) the timing game can be solved by splitting up the original timing game. The split up times are the times at which one of the payoff functions is discontinuous. In the first step, the last timing game is solved. The second last timing game is solved in the second step, where the results of the first step are used.

If no player has moved by time 1, they both receive $M(1)$. This can be interpreted as the equilibrium payoff of another game that is played if both players have not moved by time 1. Since by joint-moving the follower can at least obtain $M(t)$, it holds that

$$F(t) \geq M(t). \tag{4.43}$$

If there exist a point in time $t \in [0, 1]$ for which there is a so-called first mover advantage:

$$L(t) > F(t), \tag{4.44}$$

the timing game is called a preemption game. If for all $t \in [0,1]$:

$$F(t) > L(t), \qquad\qquad (4.45)$$

the game is called an attrition game.

The equilibrium concepts used are those introduced in Fudenberg and Tirole, 1985 for preemption games and in Hendricks et al., 1988 for attrition games. The approach of Simon, 1987a and Simon, 1987b for timing games is almost equivalent to Fudenberg and Tirole's. In Section A.2 we point out the main difference. In Chapter 8 we have to apply Simon's equilibrium concept since it is the only one that can be used for games with asymmetric players. Hendricks and Wilson, 1992 also provide an equilibrium concept for preemption games in continuous time, but they restrict themselves to strategies of only one function. However, as argued in Fudenberg and Tirole, 1985 and in Simon and Stinchcombe, 1989, strategies in preemption games must consist of two functions in order to describe mixed strategies in continuous time that are the limit of discrete time mixed strategies.

We start with describing the strategy spaces, value functions, and the equilibrium concept in Subsection A.1. After that we analyze a particular class of preemption games in Subsection A.2, and some attrition games in Subsection A.3.

A.1   STRATEGY SPACES, PAYOFF FUNCTIONS, AND EQUILIBRIUM

We use the strategy spaces that where introduced in Fudenberg and Tirole, 1985. We will first restate their definitions. They start out with defining simple continuous time strategies (such that different "types" of atoms can be distinguished), payoffs, and the Nash equilibrium. After that they extend the strategies to closed loop strategies and define the perfect equilibrium.

DEFINITION 4.1 *A simple strategy for player i in the game starting at time t is a pair of real-valued functions* $(G_i, \alpha_i) : [t,1] \times [t,1] \to [0,1] \times [0,1]$ *satisfying:*

1. G_i *is non-decreasing and right-continuous.*

2. $\alpha_i(s) > 0 \implies G_i(s) = 1.$

3. α_i *is right-differentiable.*

4. *If* $\alpha_i(s) = 0$ *and* $s = \inf(u \geq t \mid \alpha_i(u) > 0)$, *then* $\alpha_i(\cdot)$ *has positive right derivative at* s.

Condition 1 ensures that G_i is a cumulative distribution function. $G_i(s)$ is the cumulative probability that player i has moved by time s given that both players have not moved before time s. $\alpha_i(s)$ measures the intensity of atoms in the interval $[s, s + ds]$, thus condition 2 requires that if $\alpha_i(s)$ is positive then player i is sure to move by time s. The last two conditions are imposed for technical convenience. The function value $\alpha_1(s)$ ($\alpha_2(s)$) should be interpreted as the probability that firm 1 (2) chooses row (column) 1 in the matrix game of which the payoffs are depicted in Figure 4.10. Playing the game costs no time and if player 1 chooses row 2 and player 2 column 2 the game is repeated. If necessary the game will be repeated infinitely often.

	$\alpha_2(s)$ player 2 $1-\alpha_2(s)$	
$\alpha_1(s)$	$(M(s), M(s))$	$(L(s), F(s))$
player 1		
$1-\alpha_1(s)$	$(F(s), L(s))$	repeat game

Figure 4.10. Payoffs (first entry for player 1 and second entry for player 2) and strategies of matrix game played at time t.

We need some more notation in order to define the payoffs resulting from a pair of simple strategies. Define

$$\tau_i(t) = \begin{cases} 1 & \text{if } \alpha_i(s) = 0 \ \forall s \in [t, 1], \\ \inf(s \in [t, 1]| \ \alpha_i(s) > 0) & \text{otherwise.} \end{cases} \qquad (4.46)$$

At $\tau_i(t)$ the first interval of atoms in player i's strategy starts. Define

$$\tau(t) = \min(\tau_1(t), \tau_2(t)).$$

Thus in the subgame starting at time t, one of the players has ended the game for sure by time $\tau(t)$. Define

$$G_i^-(s) = \lim_{u \uparrow s} G_i(u), \qquad (4.47)$$

which is the left-hand limit of $G_i(\cdot)$ at s. The game begins at $t \geq 0$; so impose $G_i^-(t) = 0$, $i = 1, 2$. Define

$$a_i(s) = \lim_{\xi \downarrow 0}(G_i(s) - G_i(s - \xi)) = G_i(s) - G_i^-(s), \qquad (4.48)$$

which is the size of the jump in G_i at time $s\,(\geq t)$.

Define

$$V^i\left(t,(G_1,\alpha_1),(G_2,\alpha_2)\right),\qquad(4.49)$$

to be the payoff of player i in the subgame starting at time t if player j plays the simple strategy (G_j,α_j), $j=1,2$. Then payoffs are equal to

$$V^i\left(t,(G_1,\alpha_1),(G_2,\alpha_2)\right)\qquad(4.50)$$

$$=\int_{s=t}^{\tau(t)^-}\left(L\left(s\right)\left(1-G_j\left(s\right)\right)dG_i\left(s\right)+F\left(s\right)\left(1-G_i\left(s\right)\right)dG_j\left(s\right)\right)$$

$$+\sum_{s<\tau(t)}a_i\left(s\right)a_j\left(s\right)M\left(s\right)$$

$$+\left(1-G_i^-\left(\tau\left(t\right)\right)\right)\left(1-G_j^-\left(\tau\left(t\right)\right)\right)W^i\left(\tau\left(t\right),(G_1,\alpha_1),(G_2,\alpha_2)\right),$$

where, if $\tau_j\left(t\right)>\tau_i\left(t\right)$,

$$W^i\left(\tau,(G_1,\alpha_1),(G_2,\alpha_2)\right)\qquad(4.51)$$

$$=\left(\frac{a_j\left(\tau\right)}{1-G_j^-\left(\tau\right)}\right)\left(\left(1-\alpha_i\left(\tau\right)\right)F\left(\tau\right)+\alpha_i\left(\tau\right)M\left(\tau\right)\right)$$

$$+\left(\frac{1-G_j\left(\tau\right)}{1-G_j^-\left(t\right)}\right)L\left(\tau\right),$$

and, if $\tau_i\left(t\right)>\tau_j\left(t\right)$,

$$W^i\left(\tau,(G_1,\alpha_1),(G_2,\alpha_2)\right)\qquad(4.52)$$

$$=\left(\frac{a_i\left(\tau\right)}{1-G_i^-\left(\tau\right)}\right)\left(\left(1-\alpha_j\left(\tau\right)\right)L\left(\tau\right)+\alpha_j\left(\tau\right)M\left(\tau\right)\right)$$

$$+\left(\frac{1-G_i\left(\tau\right)}{1-G_i^-\left(\tau\right)}\right)F\left(\tau\right),$$

while if $\tau_i\left(t\right)=\tau_j\left(t\right)$ and $\alpha_i\left(\tau\right)=\alpha_j\left(\tau\right)=1$,

$$W^i\left(\tau,(G_1,\alpha_1),(G_2,\alpha_2)\right)=M\left(\tau\right),\qquad(4.53)$$

while if $\tau_i\left(t\right)=\tau_j\left(t\right)$ and if $0<\alpha_i\left(\tau\right)+\alpha_j\left(\tau\right)<2$,

$$W^i\left(\tau,(G_1,\alpha_1),(G_2,\alpha_2)\right)$$
$$=\frac{\alpha_i(\tau)(1-\alpha_j(\tau))L(\tau)+\alpha_j(\tau)(1-\alpha_i(\tau))F(\tau)+\alpha_i(\tau)\alpha_j(\tau)M(\tau)}{(\alpha_i(\tau)+\alpha_j(\tau)-\alpha_i(\tau)\alpha_j(\tau))},\qquad(4.54)$$

and if $\tau_i(t) = \tau_j(t)$ and if $\alpha_i(\tau) = \alpha_j(\tau) = 0$,

$$W^i(\tau, (G_1, \alpha_1), (G_2, \alpha_2)) = \frac{\alpha_i'(\tau) L(\tau) + \alpha_j'(\tau) F(\tau)}{\alpha_i'(\tau) + \alpha_j'(\tau)}. \tag{4.55}$$

The first two parts of equation (4.50) also appear in the value function of a player if the usual mixed strategy concept is used. With usual mixed strategy concept is meant the strategy concept in which a mixed strategy is represented by only one function, i.e. a distribution function. With probability $\left(1 - G_i^-(\tau(t))\right)\left(1 - G_j^-(\tau(t))\right)$ none of the players has moved by time $\tau(t)$. At least one of the cumulative distributions $G_i(\cdot)$ then jumps to one. If $\tau_j(t) > \tau_i(t) = \tau$ then the payoffs are computed as the limits of discrete time payoffs when firm i moves with probability $\alpha_i(\tau)$ at each period and firm j moves with probability $\frac{a_j(\tau)}{1 - G_j^-(\tau)}$ at the first instant and with probability zero thereafter. This corresponds to a situation in which firm j plays an isolated jump, of size $a_j(\tau)$, at time τ and firm i adopts continuously with intensity $\alpha_i(\tau)$. Firm j does not have an interval of atoms at τ because $\tau_j(t) > \tau$. If $\tau_1(t) = \tau_2(t) = \tau$, the probabilities of getting L, F, and M are computed from discrete-time limits with constant probabilities of moves $\alpha_i(\tau)$ and $\alpha_j(\tau)$. If $\alpha_i(\tau) = \alpha_j(\tau) = 0$ the payoffs are computed by a first-order Taylor expansion.

Using the value functions we can define the Nash equilibrium of a game starting at time t.

DEFINITION 4.2 *A pair of simple strategies* $\{(G_i, \alpha_i), i = 1, 2\}$ *is a Nash equilibrium of the game starting at time t (with neither player having moved yet) if each player i's strategy maximizes his payoff* $V^i(t, \cdot, \cdot)$ *holding the other player's strategy fixed.*

Next we recall Fudenberg and Tirole's definition of a closed loop strategy.

DEFINITION 4.3 *A closed loop strategy for player i is a collection of simple strategies* $\left\{\left(G_i^t(\cdot), \alpha_i^t(\cdot)\right), t \in [0, 1]\right\}$ *satisfying the intertemporal consistency conditions:*

1. $G_i^t(v) = G_i^t(u) + \left(1 - G_i^t(u)\right) G_i^u(v)$ *for* $t \le u \le v \le 1$.

2. $\alpha_i^t(v) = \alpha_i^u(v) = \alpha_i(v)$ *for* $t \le u \le v \le 1$.

The reason for the need of a whole family of strategies is that to test for perfectness, the strategies must be defined even conditional on zero-probability events. Condition 1 ensures that the family of strategies is consistent between non-zero-probability events; that is, if G_i^t puts positive weight on times from v on, then G_i^t should be consistent between time t and v. Condition 2 is a similar consistency condition. Note that we corrected the mistakes in Fudenberg and Tirole, 1985's arguments of the functions in the intertemporal consistency conditions.

DEFINITION 4.4 *A pair of closed loop strategies*

$$\left\{ \left\{ \left(G_i^t \left(\cdot \right), \alpha_i^t \left(\cdot \right) \right), t \in [0,1] \right\}, i = 1, 2 \right\}$$

is a perfect equilibrium if for every t, the pair of simple strategies

$$\left\{ \left(G_i^t \left(\cdot \right), \alpha_i^t \left(\cdot \right) \right), i = 1, 2 \right\}$$

is a Nash equilibrium.

A.2   PREEMPTION GAMES

In this section a particular class of preemption games is analyzed. We make the following additional assumptions on the value functions.

A1 $M \left(t \right)$ is continuous on $[0,1]$.

A2 $\exists T_F \in (0,1)$ such that $L \left(t \right) = F \left(t \right) = M \left(t \right) \ \forall t \in [T_F, 1]$ and $F \left(t \right) > M \left(t \right) \ \forall t \in [0, T_F)$.

A3 $F \left(t \right)$ is strictly increasing on $[0, T_F]$.

A4 $L \left(t \right) - F \left(t \right)$ is quasi concave on $[0,1]$.

As mentioned before, in a preemption game there is an incentive for the players to become the leader. Define time T_P as the first point in time at which the payoff of the leader is larger or equal than the payoff of the follower

$$T_P = \min \left(t \in [0,1] | \, L \left(t \right) \geq F \left(t \right) \right). \tag{4.56}$$

Define T_L as the point in time at which the leader curve is maximal on the interval $[0, T_F]$:

$$T_L = \arg \max_{t \in [0, T_F]} L \left(t \right). \tag{4.57}$$

LEMMA 4.1 $T_P \leq T_L$.

PROOF OF LEMMA 4.1 *Suppose $T_P > T_L$. Then since F is increasing and with the definition of T_P:*

$$L \left(T_P \right) = F \left(T_P \right) \geq F \left(T_L \right) > L \left(T_L \right),$$

which is a contradiction with the definition of T_L.   □

Define T_C as the point in time at which the joint-moving curve is at its maximum:

$$T_C = \arg \max_{t \in [0,1]} M(t).   (4.58)$$

LEMMA 4.2 $T_C \geq T_F$.

PROOF OF LEMMA 4.2 *Since F is increasing on $[0, T_F]$, $M(t) \leq F(t)$ for $t \in [0,1]$, and $M(T_F) = F(T_F)$, it holds that $M(s) < M(T_F)$ for $s \in [0, T_F]$, so that $T_C \geq T_F$.*   □

We distinguish two cases. In the first case it holds that $L(T_L) > M(T_C)$ and in the second case $L(T_L) \leq M(T_C)$. An example of the first (second) case is depicted in Figure 4.2 (4.3).

CASE 1: $L(T_L) > M(T_C)$

Both players would like to move at time T_L, since that would give them the largest possible payoff. Joint-movement is not optimal since $F(t) > M(t)$ for $t \in [0, T_L)$. Knowing this, one player, say player 1, will try to preempt player 2 by stopping at time $T_L - \varepsilon$, but then player 2 tries to preempt player 1 by moving at time $T_L - 2\varepsilon$ and so forth and so on. This preemption process stops at time T_P. Either the leader and follower curves are equal at time T_P, which is called rent equalization, or there is no rent equalization and $T_P = 0$.

Fudenberg and Tirole, 1985 prove that if there is rent equalization the probability of a mistake, that is both players stopping at time T_P, is zero. If there is no rent equalization, the probability of a mistake is positive. The equilibrium strategy for each player is given by (cf. Fudenberg and Tirole, 1985):

$$G^t(s) = \begin{cases} 0 & s \in [t, T_P), \\ 1 & s \in [T_P, 1], \end{cases}   (4.59)$$

$$\alpha(s) = \begin{cases} 0 & s \in [t, T_P), \\ \frac{L(s)-F(s)}{L(s)-M(s)} & s \in (T_P, T_C), \\ 1 & s \in [T_C, 1]. \end{cases}   (4.60)$$

Let us derive $\alpha(s)$ for $s \in (T_P, T_C)$. Suppress the time arguments and denote the payoff of a player i by $P_i(\alpha_i, \alpha_j)$, with $i, j \in \{1, 2\}$ and $i \neq j$. From Figure 4.10 it follows that

$$P_i(\alpha_i, \alpha_j) = \alpha_i \alpha_j M + \alpha_i (1 - \alpha_j) L + (1 - \alpha_i) \alpha_j F$$
$$+ (1 - \alpha_i)(1 - \alpha_j) P_i(\alpha_i, \alpha_j).   (4.61)$$

Rewriting gives

$$P_i\left(\alpha_i, \alpha_j\right) = \frac{\alpha_i \alpha_j M + \alpha_i \left(1 - \alpha_j\right) L + \left(1 - \alpha_i\right) \alpha_j F}{1 - \left(1 - \alpha_i\right) \left(1 - \alpha_j\right)}. \tag{4.62}$$

To find the optimal value for α_i we differentiate (4.62) with respect to α_i and put this expression equal to zero. This eventually leads to the following equality:

$$\frac{\partial P_i\left(\alpha_i, \alpha_j\right)}{\partial \alpha_i} = \frac{\alpha_j \left(\left(1 - \alpha_j\right) L - F + \alpha_j M\right)}{\left(1 - \left(1 - \alpha_i\right) \left(1 - \alpha_j\right)\right)^2} = 0. \tag{4.63}$$

It is easily verified that $\frac{\partial^2 P_i(\alpha_i, \alpha_j)}{\partial \alpha_i^2} < 0$, so that satisfying (4.63) indeed leads to a maximum value of the player. Since we only consider symmetrical strategies we impose that

$$\alpha_i = \alpha_j = \alpha. \tag{4.64}$$

Combining (4.63) and (4.64) leads to the following optimal value for α:

$$\alpha = \frac{L - F}{L - M}. \tag{4.65}$$

Consider a subgame that starts at time s for which $L\left(s\right) \geq F\left(s\right)$. The probability that a player stops the game at time s, $\Pr\left(\text{one}|\, s\right)$, equals

$$\Pr\left(\text{one}|\, s\right) \;=\; \alpha\left(s\right)\left(1 - \alpha\left(s\right)\right) + \left(1 - \alpha\left(s\right)\right)\left(1 - \alpha\left(s\right)\right)\Pr\left(\text{one}|\, s\right),$$

so that

$$\Pr\left(\text{one}|\, s\right) \;=\; \frac{1 - \alpha\left(s\right)}{2 - \alpha\left(s\right)}, \tag{4.66}$$

and the probability that both players stop the game at s, $\Pr\left(\text{two}|\, s\right)$, equals

$$\Pr\left(\text{two}|\, s\right) \;=\; \alpha\left(s\right)\alpha\left(s\right) + \left(1 - \alpha\left(s\right)\right)\left(1 - \alpha\left(s\right)\right)\Pr\left(\text{two}|\, s\right),$$

so that

$$\Pr\left(\text{two}|\, s\right) \;=\; \frac{\alpha\left(s\right)}{2 - \alpha\left(s\right)}. \tag{4.67}$$

Thus each player stops the game itself with probability $\frac{1 - \alpha(s)}{2 - \alpha(s)}$ and with probability $\frac{\alpha(s)}{2 - \alpha(s)}$ both players stop the game. If there is no rent equalization, i.e. $L \neq F$, by (4.65) and (4.67) the probability of a mistake is positive.

If at time $t = T_P$ there is rent equalization, i.e. $L = F$, by (4.65) it holds that $\alpha(T_P) = 0$ and the probabilities are equal to

$$\Pr(\text{one}|\, s) \;=\; \frac{1}{2}, \tag{4.68}$$

$$\Pr(\text{two}|\, s) \;=\; 0. \tag{4.69}$$

By (4.68) we get that the probability that one player becomes leader at time T_P is equal for both players (one-half). Moreover, from equation (4.69) it follows that when there is rent equalization, i.e. $L = F$, the probability of a mistake, that is both players stopping at time T_P and thus gaining the lowest possible payoff M, is zero. Mathematically this means that $\alpha_i(T_P) = 0$ for both players ($i = 1, 2$).

The first mover advantage ($L(s) > F(s)$) results in equilibrium strategies in which both players take a positive chance of making a mistake in order to get the leader payoff. Substitution of equations (4.59) and (4.60) into equation (4.50) shows that a player sets his intensity $\alpha(\cdot)$ such that his expected value equals the follower value: the expected payoff $E\left[V^i(t, \cdot, \cdot)\right]$ for each player $i\,(= 1, 2)$ of the subgame starting at some time $t \in [0, 1]$ equals:

$$E\left[V^i(t, \cdot, \cdot)\right] = \left\{ \begin{array}{ll} F(T_P) & t \in [0, T_P], \\ F(t) & t \in (T_P, 1]. \end{array} \right. \tag{4.70}$$

We summarize in the following proposition. For a formal proof we refer to Fudenberg and Tirole, 1985.

PROPOSITION 4.7 *The equilibrium strategies for the preemption game that satisfies assumptions A1-A4 and for which $L(T_L) > M(T_C)$ are given by equations (4.59) and (4.60). The expected payoff for each player is given by equation (4.70).*

CASE 2: $L(T_L) \leq M(T_C)$

In this scenario there are multiple equilibria. The equilibria can be divided into two types. The first type is the preemption equilibrium defined in the previous subsection and the second type is a so-called joint-movement equilibrium.

Define

$$T_S = \min\left(t \geq T_F |\, M(t) = L(T_L)\right). \tag{4.71}$$

There are an infinite number of type 2 equilibria. Each equilibrium is characterized by its movement date u, where $u \in [T_S, T_C]$. Equilibrium strategies are given by

$$G^t(s) = \begin{cases} 0 & s \in [t, u), \\ 1 & s \in [u, 1], \end{cases} \qquad (4.72)$$

$$\alpha(s) = \begin{cases} 0 & s \in [t, u), \\ 1 & s \in [u, 1]. \end{cases} \qquad (4.73)$$

Fudenberg and Tirole, 1985 argue that the Pareto-superior joint moving equilibrium, both moving at time T_C, is the most reasonable outcome of the game.

PROPOSITION 4.8 *The preemption game that satisfies assumptions A1-A4 and for which it holds that $L(T_L) \leq M(T_C)$ has two types of equilibrium strategies. The first type is given by equations (4.59)-(4.60) and the second type by equations (4.72)-(4.73). The most reasonable outcome of the game is joint-movement at time T_C.*

Now we are in a position to point out the main difference with the approach above and the one developed in Simon, 1987a and Simon, 1987b. In case 2 the most reasonable outcome is the only equilibrium if Simon's equilibrium concept is used.

A.3   ATTRITION GAMES

In this section we analyze timing games that satisfy the following assumptions:

A5 $F(t) > M(t)$ for $t \in [0, 1)$.

A6 $L(t)$ is strictly decreasing for $t \in [0, 1)$.

Since, contrary to a preemption game, both players do not want to take any chances of making a mistake in a war of attrition, the only important joint-movement value is $M(1)$. In other words, the outcome is not influenced by the shape of the joint-movement curve before time 1 as long as it is below the follower value. This implies for the equilibrium strategies that $\alpha(s) = 0$ for $s \in [0, 1)$.

We distinguish two different cases. In the first case $L(t) > M(1)$ for $t \in [0, 1]$ and in the second case $L(t^*) = M(1)$ for some unique $t^* \in [0, 1]$. In the left (right) panel of Figure 4.11 the leader, follower, and joint-moving curves are plotted for the first (second) case.

CASE 1: $L(T) > M(1)$ FOR $T \in [0, 1]$

Hendricks et al., 1988 show that there is no symmetric equilibrium for this game. There are two asymmetric equilibria. In the first one player 1

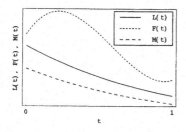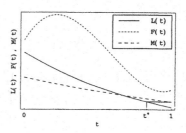

Figure 4.11. Payoff curves in case 1 (left panel) and case 2 (right panel).

stops at time 0 and in the second one player 2 stops at time 0. Although this result is unsatisfactory it is understandable. Consider a symmetric strategy. Since L is decreasing and $M(1) < L(1)$ both players will try not to reach time 1 without one of them having stopped the game before. But, to do so they have to apply a strategy with $G_i^t(s) = 1$ for $s < 1$. But if a player knows that the other player stops the game before time 1 with probability 1, his optimal strategy is not to stop at all, implying that $G^t(s) = 0$ for $s \in [t, 1]$. Thus there does not exist an equilibrium with symmetric strategies.

PROPOSITION 4.9 *In an attrition game that satisfies assumptions A5-A6 and for which $M(t) < L(t)$ for $t \in [0,1]$ there does not exist an equilibrium with symmetric strategies.*

CASE 2: $L(T^*) = M(1)$ FOR SOME UNIQUE $T^* \in [0,1]$

For this scenario there is an equilibrium with symmetric strategies. Hendricks et al., 1988 show that the equilibrium strategies are given by, if $t < t^*$:

$$
G^t(s) = \begin{cases}
1 - e^{\int_{v=t}^{s} \frac{dL(v)}{(F(v)-L(v))}} & s \in [t, t^*], \\
1 - e^{\int_{v=t}^{t^*} \frac{dL(v)}{(F(v)-L(v))}} & s \in (t^*, 1), \\
1 & s = 1,
\end{cases}
\tag{4.74}
$$

$$
a^t(1) = e^{\int_{v=t}^{t^*} \frac{dL(v)}{(F(v)-L(v))}}, \tag{4.75}
$$

$$
\alpha(s) = \begin{cases}
0 & s < 1, \\
1 & s = 1,
\end{cases}
\tag{4.76}
$$

and if $t \geq t^*$,

$$G^t(s) = \begin{cases} 0 & s \in [t,1), \\ 1 & s = 1, \end{cases} \tag{4.77}$$

$$a^t(1) = 1, \tag{4.78}$$

$$\alpha(s) = \begin{cases} 0 & s \in [t,1), \\ 1 & s = 1. \end{cases} \tag{4.79}$$

Equations (4.75) and (4.78) imply that there is a discontinuous jump in G^t at time 1. The same argument as in the previous subsection applies here. If G would be equal to 1 before time 1 the other player would be better of by setting his $G(t) = 0$ for all t less than 1. Thus the positive probability of reaching time 1 while neither of the players has moved, $\left(a^t(1)\right)^2$, enforces each player to stop the game before time t^* with positive probability. Since in this scenario the players take chances of getting the terminal payoff, contrary to the previous subsection, an equilibrium with symmetric strategies exists. Note that the existence of t^* is needed for the existence of the symmetric equilibrium.

PROPOSITION 4.10 *In an attrition game that satisfies assumptions A5-A6 and for which $L(t^*) = M(1)$ for some unique $t^* \in [0,1]$ there exists an equilibrium with symmetric strategies, given by equations (4.74)-(4.79).*

B.   LEMMA AND PROOFS

The following lemma disproves a claim by Stenbacka and Tombak, 1994.

LEMMA 4.3 *There does not exist a unique date $t_S^{**} = t_L = t_F$ which satisfies:*

$$\begin{cases} \dfrac{\lambda(\pi(\theta_1,\theta_0)-\pi(\theta_0,\theta_0))}{(r+\lambda)} + \dfrac{\lambda\frac{\partial t_F}{\partial t_L}e^{-r(t_F-t_L)}(\pi(\theta_1,\theta_1)-\pi(\theta_1,\theta_0))}{(r+\lambda)} \\ \quad + \dfrac{\left((r+\lambda)\frac{\partial t_F}{\partial t_L}-\lambda\right)\lambda e^{-(r+\lambda)(t_F-t_L)}(\pi(\theta_0,\theta_1)-\pi(\theta_1,\theta_1)-\pi(\theta_0,\theta_0)+\pi(\theta_1,\theta_0))}{(r+\lambda)(r+2\lambda)} \\ \quad + \left.\dfrac{\partial I(t)}{\partial t}\right|_{t=t_L} - rI(t_L) = 0, \\[2ex] \dfrac{\lambda(\pi(\theta_1,\theta_1)-\pi(\theta_0,\theta_1))}{(r+\lambda)} + \dfrac{\lambda e^{-\lambda(t_F-t_L)}(\pi(\theta_0,\theta_1)-\pi(\theta_1,\theta_1)-\pi(\theta_0,\theta_0)+\pi(\theta_1,\theta_0))}{(r+2\lambda)} \\ \quad + \left.\dfrac{\partial I(t)}{\partial t}\right|_{t=t_F} - rI(t_F) = 0. \end{cases}$$

$$\tag{4.80}$$

PROOF OF LEMMA 4.3 *Set $t_L = t_F$ in (4.80) and substitute the second equation in the first equation:*

$$\frac{\lambda(\pi(\theta_1,\theta_0)-\pi(\theta_0,\theta_0))}{(r+\lambda)} + \frac{\lambda \left.\frac{\partial t_F}{\partial t_L}\right|_{t_L=t_F}(\pi(\theta_1,\theta_1)-\pi(\theta_1,\theta_0))}{(r+\lambda)}$$

$$+ \frac{\left((r+\lambda)\left.\frac{\partial t_F}{\partial t_L}\right|_{t_L=t_F} -\lambda\right)\lambda(\pi(\theta_0,\theta_1)-\pi(\theta_1,\theta_1)-\pi(\theta_0,\theta_0)+\pi(\theta_1,\theta_0))}{(r+\lambda)(r+2\lambda)}$$

$$= \frac{\lambda(\pi(\theta_1,\theta_1)-\pi(\theta_0,\theta_1))}{(r+\lambda)} + \frac{\lambda(\pi(\theta_0,\theta_1)-\pi(\theta_1,\theta_1)-\pi(\theta_0,\theta_0)+\pi(\theta_1,\theta_0))}{(r+2\lambda)}. \quad (4.81)$$

Rearranging gives

$$\left.\frac{\partial t_F}{\partial t_L}\right|_{t_L=t_F}\left(\frac{\lambda(\pi(\theta_1,\theta_1)-\pi(\theta_1,\theta_0))}{(r+\lambda)} + \frac{\lambda(\pi(\theta_0,\theta_1)-\pi(\theta_1,\theta_1)-\pi(\theta_0,\theta_0)+\pi(\theta_1,\theta_0))}{(r+2\lambda)}\right) = 0. \quad (4.82)$$

So that at least one of the following equations has to hold:

$$\left.\frac{\partial t_F}{\partial t_L}\right|_{t_L=t_F} = 0, \quad (4.83)$$

$$\frac{\lambda(\pi(\theta_1,\theta_1)-\pi(\theta_1,\theta_0))}{(r+\lambda)} + \frac{\lambda(\pi(\theta_0,\theta_1)-\pi(\theta_1,\theta_1)-\pi(\theta_0,\theta_0)+\pi(\theta_1,\theta_0))}{(r+2\lambda)} = 0. \quad (4.84)$$

Stenbacka and Tombak, 1994 derived the following equation for $\frac{\partial t_F}{\partial t_L}$:

$$\frac{\frac{\lambda^2 e^{-\lambda(t_F-t_L)}(\pi(\theta_1,\theta_0)-\pi(\theta_1,\theta_1)-\pi(\theta_0,\theta_0)+\pi(\theta_0,\theta_1))}{(r+2\lambda)}}{\frac{\lambda^2 e^{-\lambda(t_F-t_L)}(\pi(\theta_1,\theta_0)-\pi(\theta_1,\theta_1)-\pi(\theta_0,\theta_0)+\pi(\theta_0,\theta_1))}{(r+2\lambda)}+r\left.\frac{\partial I(t)}{\partial t}\right|_{t=t_F}-\left.\frac{\partial^2 I(t)}{\partial t^2}\right|_{t=t_F}}. \quad (4.85)$$

Due to equations (4.2) and (4.85), equation (4.83) can not hold. Rewriting equation (4.84) gives

$$\frac{\lambda^2\left(\pi\left(\theta_1,\theta_1\right)-\pi\left(\theta_1,\theta_0\right)\right)+\lambda\left(r+\lambda\right)\left(\pi\left(\theta_0,\theta_1\right)-\pi\left(\theta_0,\theta_0\right)\right)}{\left(r+\lambda\right)\left(r+2\lambda\right)} = 0. \quad (4.86)$$

Due to equation (4.3) this equation can not hold. □

PROOF OF PROPOSITION 4.5 *The first-order condition for interior solutions for the leader in the feedback game is*

$$\frac{\lambda\left(\pi\left(\theta_0,\theta_0\right)-\pi\left(\theta_1,\theta_0\right)\right)}{(r+\lambda)}-\frac{\lambda\frac{\partial t_F}{\partial t_L}e^{-r(t_F-T_L)}\left(\pi\left(\theta_1,\theta_1\right)-\pi\left(\theta_1,\theta_0\right)\right)}{(r+\lambda)}$$

$$+\frac{\lambda\frac{\partial t_F}{\partial t_L}e^{-(r+\lambda)(t_F-T_L)}\left(\pi\left(\theta_1,\theta_1\right)-\pi\left(\theta_1,\theta_0\right)\right)}{(r+2\lambda)}$$

$$-\frac{\lambda^2 e^{-(r+\lambda)(t_F-T_L)}\left(\pi\left(\theta_1,\theta_1\right)-\pi\left(\theta_1,\theta_0\right)\right)}{(r+\lambda)(r+2\lambda)}$$

$$+\frac{\lambda\left(\lambda-(r+\lambda)\frac{\partial t_F}{\partial t_L}\right)e^{-(r+\lambda)(t_F-T_L)}\left(\pi\left(\theta_0,\theta_1\right)-\pi\left(\theta_0,\theta_0\right)\right)}{(r+\lambda)(r+2\lambda)}$$

$$+\ rI\left(T_L\right)-\frac{\partial I\left(t\right)}{\partial t}\Bigg|_{t=T_L}=0. \tag{4.87}$$

Note that we correct for the (two) sign mistakes in Stenbacka and Tombak's equation (11). In the open loop case the first order condition is given by the following equation:

$$\frac{\lambda\left(\pi\left(\theta_0,\theta_0\right)-\pi\left(\theta_1,\theta_0\right)\right)}{(r+\lambda)}-\frac{\lambda^2 e^{-(r+\lambda)\left(t_F-t_L^*\right)}\left(\pi\left(\theta_1,\theta_1\right)-\pi\left(\theta_1,\theta_0\right)\right)}{(r+\lambda)(r+2\lambda)}$$

$$+\frac{\lambda^2 e^{-(r+\lambda)\left(t_F-t_L^*\right)}\left(\pi\left(\theta_0,\theta_1\right)-\pi\left(\theta_0,\theta_0\right)\right)}{(r+\lambda)(r+2\lambda)}+rI\left(t_L^*\right)-\frac{\partial I\left(t\right)}{\partial t}\Bigg|_{t=t_L^*}$$

$$=0. \tag{4.88}$$

Now, define the following function:

$$f\left(t_L\right)\ =\ \frac{\lambda\left(\pi\left(\theta_0,\theta_0\right)-\pi\left(\theta_1,\theta_0\right)\right)}{(r+\lambda)}$$

$$-\frac{\lambda^2 e^{-(r+\lambda)(t_F-t_L)}\left(\pi\left(\theta_1,\theta_1\right)-\pi\left(\theta_1,\theta_0\right)\right)}{(r+\lambda)(r+2\lambda)}$$

$$+\frac{\lambda^2 e^{-(r+\lambda)(t_F-t_L)}\left(\pi\left(\theta_0,\theta_1\right)-\pi\left(\theta_0,\theta_0\right)\right)}{(r+\lambda)(r+2\lambda)}+rI\left(t_L\right)-$$

$$\frac{\partial I\left(t\right)}{\partial t}\Bigg|_{t=t_L}. \tag{4.89}$$

From (4.88) and (4.89) it follows that

$$f\left(t_L^*\right)=0. \tag{4.90}$$

Provided that the second order condition holds (cf. p. 409 of Stenbacka and Tombak, 1994), it can be shown that

$$\frac{\partial f(t_L)}{\partial t_L} < 0. \tag{4.91}$$

Furthermore, observe that (4.87) can be written into

$$f(T_L) - \frac{\lambda \frac{\partial t_F}{\partial t_L} e^{-r(t_F - T_L)} (\pi(\theta_1, \theta_1) - \pi(\theta_1, \theta_0))}{(r + \lambda)}$$

$$+ \frac{\lambda \frac{\partial t_F}{\partial t_L} e^{-(r+\lambda)(t_F - T_L)} (\pi(\theta_1, \theta_1) - \pi(\theta_1, \theta_0))}{(r + 2\lambda)}$$

$$- \frac{\lambda \frac{\partial t_F}{\partial t_L} e^{-(r+\lambda)(t_F - T_L)} (\pi(\theta_0, \theta_1) - \pi(\theta_0, \theta_0))}{(r + 2\lambda)} = 0. \tag{4.92}$$

Since the sum of the three terms is negative, it follows from (4.92) that

$$f(T_L) > 0. \tag{4.93}$$

Now (4.91), (4.92), and (4.93) imply that $T_L < t_L^*$. □

Chapter 5

TWO NEW TECHNOLOGIES

1. INTRODUCTION

One of the features of the models concerning the investment of new technologies considered in the previous chapter is that only one new technology was available. The availability of more consecutive new technologies complicates the technology investment decision considerably, since every time the firm evaluates an investment in a new technology it has to take into account that at a later point of time a more efficient technology will be invented.

The aim of this chapter is to provide a first step in analyzing the problem of when a firm could adopt an existing technology knowing that a better technology will become available later, while it has to fight for a market share with an identical firm on the output market. Two technologies are considered: an existing one which can be adopted immediately, and a new one which is more efficient and enters the input market at a known future date. Learning is incorporated in the sense that it is less costly to adopt and successfully implement the new technology if it has adopted the current technology before. As such this framework is taken from Grenadier and Weiss, 1997. In that paper the future date at which the new technology becomes available is uncertain and only one firm is considered. So, compared to Grenadier and Weiss, 1997 we exchange the uncertainty for competition on the output market. In this way we are able to identify the strategic aspects of this problem.

Two scenarios are worked out in detail: one where the new technology is cheap, and one where the new technology is so expensive that it is not optimal for both firms to produce with the new technology. In the latter case we show that on a particular time interval it is optimal for

107

one firm to invest right away in the current technology while the other firm waits with investment in order to adopt the new technology as soon as it becomes available. Which firm will do better depends on the comparison between the temporary monopoly profits gained by the first investor before the new technology arrives, versus the higher revenue the other firm obtains after the arrival date of the new technology due to the fact that it produces with a more efficient technology. In case the monopoly profits are outweighed by the higher revenue associated with the new technology, second mover advantages arise. To our knowledge, the way second mover advantages are caused here has not occurred in the literature yet. Different second mover advantages have been found by Hendricks, 1992 and Dutta et al., 1995. In Hendricks, 1992 they are caused by ex ante uncertainty in the profitability of adoption (see also Hoppe, 2000), while in Dutta et al., 1995 the quality of the product improves over time. These second mover advantages may lead to a better understanding of the fact that from empirical studies it could not always be concluded that early entrants perform better than later entrants. Apart from the numerous studies that found persistent market-share advantages to first entrants, there are many examples of pioneering firms that did not survive the competition of later entrants. Dutta et al., 1995 mention the case of EMI, which developed the first CT scanner but lost its market place because it lacked a technological infrastructure and marketing base in the medical field.

The contents of this chapter is as follows. The model is introduced in Section 2. In Section 3 the solution procedure is explained and optimal investment strategies are analyzed in detail for two specific scenarios. Section 4 concludes.

2.   THE MODEL

The model is based on Grenadier and Weiss, 1997, but here a duopoly with two identical risk-neutral and value maximizing firms is considered, while in Grenadier and Weiss, 1997 the analysis is focussed on a single firm. To produce goods the firms need to acquire a certain technology. Initially, at time $t = 0$, they can invest in a current technology, of which the efficiency is denoted by $\theta_1 (> 0)$. At time $t = T (\geq 0)$, a new and more efficient technology becomes available for adoption, with efficiency $\theta_2 (> \theta_1)$. In our analysis time T is assumed to be known beforehand (contrary to Grenadier and Weiss, 1997, where T depends on the realization of a Wiener process that governs the state of technological knowledge). When a firm does not produce we denote this by $\theta_0 = 0$. The firm's profits per unit of time, while it produces with technology θ_i and the other firm with technology θ_j, are equal to $\pi(\theta_i, \theta_j)$, with

$i, j \in \{0, 1, 2\}$. We assume that for $j \in \{0, 1, 2\}$:

$$\pi(\theta_0, \theta_j) = 0, \tag{5.1}$$

and for $i \in \{1, 2\}$ and $j \in \{0, 1, 2\}$:

$$\pi(\theta_i, \theta_j) > 0. \tag{5.2}$$

If P_{ij} denotes the value of the firm while this firm itself produces with technology θ_i forever and the other firm with technology θ_j forever, it holds that

$$P_{ij} = \int_{s=0}^{\infty} \pi(\theta_i, \theta_j) e^{-rs} ds = \frac{\pi(\theta_i, \theta_j)}{r}, \tag{5.3}$$

where $r \, (> 0)$ is the constant discount rate.

If the new technology is not available for adoption yet, i.e. $t < T$, the firm has the possibility to invest in the current technology, where the investment expenditure equals $C_e \, (> 0)$. Then the firm's payoff is $P_{1j} - C_e$, where $j = 1$ when the other firm is producing with technology θ_1 and $j = 0$ when the other firm refrains from producing.

From time T onwards the firm can choose to adopt the new technology. If the firm has invested in the current technology before, it may replace this technology by the new one. Then the payoff of investing in the new technology is $P_{2i} - P_{1j} - C_u$, where $C_u \, (> 0)$ stands for the cost of upgrading. Note that in the formulation of the payoff it is taken into account that the other firm can change its technology too at time T.

If the firm adopts the new technology without having invested in the current technology before, the payoff of this investment is $P_{2i} - C_l$, with $C_l > 0$. At the moment the new technology arrives the demand for the current technology will fall so that it makes sense that the acquisition cost of the current technology will fall too. This makes that if the firm did not buy the current technology before, it may become profitable to adopt this technology after time T. The payoff of this transaction is $P_{1i} - C_d$, with $C_d > 0$.

Concerning the levels of the different cost parameters we impose that

$$C_u < C_l < C_e + C_u, \tag{5.4}$$

$$C_d < C_e. \tag{5.5}$$

The first inequality in (5.4) denotes the learning effect in the sense that it is less costly to adopt and successfully implement the new technology if the firm already produces goods. The second inequality assures that no arbitrage is possible, i.e. it is always more costly to immediately start producing with the current technology and replacing it later by the new one, than to refrain from production initially in order to wait for the new technology to arrive.

The cost of a particular technology falls over time, because (1) it becomes old-fashioned, (2) the firms that are most eager to buy the technology have already bought it so that technology suppliers have to drop their price in order to find additional buyers, and (3) due to learning by doing the technology supplier can produce the technology in a cheaper way. For these reasons we assume here that technology investments are irreversible.

Next, we specify how the profit streams are related to each other. Note that from (5.3) it can be obtained that profits per unit of time $\pi(\theta_i, \theta_j)$ are related to each other in the same way as the discounted profit streams P_{ij}. First, it holds that when the firm produces with a given technology the highest profit it can obtain is the monopoly profit, which the firm receives if the other firm does not produce. Second, its profits will be lowest when the other firm is a strong competitor in the sense that it produces in the most efficient manner by using the modern technology. This leads to

$$P_{i0} > P_{i1} > P_{i2} \quad \text{for } i \in \{1, 2\}. \tag{5.6}$$

Furthermore, by upgrading its technology, thus exchanging the current technology for the new technology, the firm gains most the less competitive the other firm is. Of course, since the new technology is more efficient, the profit stream always increases due to this exchange. Mathematically, this can be expressed as

$$P_{21} - P_{11} > P_{22} - P_{12} > 0. \tag{5.7}$$

If the other firm produces more efficiently the home firm has a lower market share and thus produces less products. Therefore, adopting a more efficient technology leads to an efficiency gain for producing less products so that the revenue increase due to this technology investment is lower than when the other firm produces with the current technology.

Finally, in order to limit the number of possible cases, we focus on the scenarios where for each technology investment the discounted future profit stream exceeds the immediate expenditure. Due to (5.4), (5.5), and (5.6) it can be concluded that this is assured by

$$P_{12} > C_e \quad \text{and} \quad P_{22} > C_l. \tag{5.8}$$

3.   SOLUTION PROCEDURE

3.1   CANDIDATE STRATEGIES FOR OPTIMALITY

Since for every technology investment the discounted future profit stream exceeds the immediate cost expenditure (cf. (5.8)), it is optimal for each firm to invest at least once. This implies that, given that we are at time $t = 0$, each firm has four candidate strategies for optimality (cf. Grenadier and Weiss, 1997). Note that due to discounting the firms will either invest at time $t = 0$ or at time $t = T$. Strategic interactions can enforce the firms to invest at other points in time. This especially happens in the case with second mover advantage (see Section 3.4).

The first strategy is called the *Compulsive strategy*. Here the firm invests right away in the current technology, and replaces this current technology by the new one as soon as the latter becomes available. The payoff of the Compulsive strategy equals

$$P_{1j} - C_e + e^{-rT} \left(P_{2i} - P_{1j} - C_u \right). \tag{5.9}$$

In (5.9), as well as below in (5.10), (5.11), and (5.12), the other firm produces with technology θ_j before time T and with technology θ_i after time T, with $j \leq i$. Of course, if j or i equals 0 it is meant that the other firm does not produce at all.

The second strategy is the *Buy and Hold strategy* by which it is meant that the firm invests right away in the current strategy and keeps on producing with it forever. The firm's payoff then equals

$$P_{1j} - C_e + e^{-rT} \left(P_{1i} - P_{1j} \right). \tag{5.10}$$

The third strategy is the *Leapfrog strategy*. Then the firm waits for the new technology to arrive and adopts it then. The payoff of this strategy is

$$e^{-rT} \left(P_{2i} - C_l \right). \tag{5.11}$$

The fourth strategy is to wait for the new technology to arrive, and at that moment invest in the current technology, which then can be bought against a cheaper price. The payoff of this so-called *Laggard strategy* then equals

$$e^{-rT} \left(P_{1i} - C_d \right). \tag{5.12}$$

3.2   EQUILIBRIUM STRATEGIES

The equilibrium strategies of both firms depend on the scenarios in which they have to operate. It holds that in some scenarios upgrading

Table 5.1.   Possible scenarios and their specifications.

Scenario	Specifications							
1	$P_{22} - P_{12}$	$<$	$P_{21} - P_{11}$	$\leq$	$C_l - C_d, C_u$			
2	$P_{22} - P_{12}$	$\leq$	$C_l - C_d$	$<$	$P_{21} - P_{11}$	$\leq$	C_u	
3	$P_{22} - P_{12}$	$\leq$	C_u	$<$	$P_{21} - P_{11}$	$\leq$	$C_l - C_d$	
4	$P_{22} - P_{12}$	$\leq$	$C_l - C_d, C_u$	$<$	$P_{21} - P_{11}$			
5	$C_l - C_d$	$<$	$P_{22} - P_{12}$	$<$	$P_{21} - P_{11}$	$\leq$	C_u	
6	$C_l - C_d$	$<$	$P_{22} - P_{12}$	$\leq$	C_u	$<$	$P_{21} - P_{11}$	
7	C_u	$<$	$P_{22} - P_{12}$	$<$	$P_{21} - P_{11}$	$\leq$	$C_l - C_d$	
8	C_u	$<$	$P_{22} - P_{12}$	$\leq$	$C_l - C_d$	$<$	$P_{21} - P_{11}$	
9	$C_l - C_d, C_u$	$<$	$P_{22} - P_{12}$	$<$	$P_{21} - P_{11}$			

is optimal, implying that the payoff of the Compulsive strategy exceeds the payoff of the Buy and Hold strategy, while in other ones it is not. Another factor that distinguishes the different scenarios are the payoffs of the Leapfrog and the Laggard strategies: the Leapfrog payoff can exceed the Laggard payoff but it can be the other way round too. Here it also has to be taken into account that the ranking of the payoffs depends on what the other firm is doing: producing with the current technology or with the new one. On the other hand, all scenarios have in common that when the firm exchanges the current technology for the new technology, it gains more when the other firm produces with the current technology instead of the new technology (cf. (5.7)).

In Table 5.1 all possible scenarios are listed. In total there are nine scenarios, each giving a different solution. It would lead to using up too much space and unnecessary repetitions if in the sequel we would study all these solutions. Instead we describe two of these solutions in detail in the next subsections. In order to still cover many different aspects of optimal technology investments we choose rather opposite scenarios. In the first case it is relatively cheap to acquire the new technology, while in the second case the new technology is expensive and the learning effect is negligible.

In both subsections we start out by analyzing the case of exogenous firm roles, i.e., despite the fact that both firms are identical one of them is given the leader role beforehand. This implies that only this firm is allowed to invest first. The other firm is the follower, which can choose between investing at the same time as the leader resulting in joint adoption, or investing later. The resulting solution is taken as a starting point to consider the more realistic case of endogenous firm roles, meaning that beforehand it is not known which firm will be the leader.

3.3   EQUILIBRIUM STRATEGIES IF THE NEW TECHNOLOGY IS CHEAP

The scenario we have in mind here is number 9 in Table 5.1, from which it can be obtained that it must hold that

$$P_{21} - P_{11} > P_{22} - P_{12} > \max\left(C_u, C_l - C_d\right). \qquad (5.13)$$

Due to (5.9) and (5.10) we can conclude that under (5.11) Compulsive dominates Buy and Hold, while (5.11) and (5.12) imply that Leapfrog dominates Laggard.

EXOGENOUS FIRM ROLES

Despite of the fact that both firms are identical one of them gets the leader role beforehand so that the other firm is the follower. Straightforward calculations lead to the equilibrium strategies that are presented in Table 5.2. Concerning the notation, $T_{xy,z}$ means that: (i) if the second technology arrives exactly at this point of time a firm is indifferent between strategy x and y, given that the other firm performs strategy z, and (ii) if the second technology arrives before (after) time $T_{xy,z}$ the firm prefers strategy x (y), given that the other firm uses strategy z. Here the names of the strategies are abbreviated (C: Compulsive, Le: Leapfrog, La: Laggard, and B: Buy and Hold). So, if for instance the arrival date of the second technology $T \in [0, T_{LeC,Le}]$, according to Table 5.2 both firms should follow a Leapfrog strategy. With equations (5.9) and (5.11) we derived that

$$T_{LeC,Le} = \frac{1}{r} \log\left(\frac{P_{10} - C_l + C_u}{P_{10} - C_e}\right), \qquad (5.14)$$

$$T_{LeC,C} = \frac{1}{r} \log\left(\frac{P_{11} - C_l + C_u}{P_{11} - C_e}\right). \qquad (5.15)$$

Equations (5.4) and (5.6) imply that

$$0 < T_{LeC,Le} < T_{LeC,C}. \qquad (5.16)$$

Since in scenario 9 the new technology is attractive, only those strategies occur under which this technology will be bought: Compulsive and Leapfrog. On the first interval $[0, T_{LeC,Le}]$ the arrival of the new technology is that near that for both firms it is optimal to wait with investment

Table 5.2. Equilibrium strategies in scenario 9 as function of T, the arrival time of the second technology, when the firm roles are assigned exogenously.

T interval	Leader	Follower
$[0, T_{LeC,Le}]$	Leapfrog	Leapfrog
$(T_{LeC,Le}, T_{LeC,C}]$	Compulsive	Leapfrog
$(T_{LeC,C}, \infty)$	Compulsive	Compulsive

until the new technology becomes available. This explains the occurrence of the Leapfrog strategy on this interval.

On the time interval $(T_{LeC,Le}, T_{LeC,C}]$ T is a bit further away, which implies that, given that the leader announces a Compulsive strategy, the follower will prefer Leapfrog, i.e. refrain from immediate investment in order to wait for the new technology to arrive. The explanation is that the time interval in which the current technology will be used is too short to make investing in the current technology profitable. Also the learning effect, i.e. implementing the new technology is cheaper when the firm has already production experience due to using the current technology, cannot make up for this (cf. (5.4)). But, given that the follower will not adopt the current technology so that it will not produce before time T, by investing in the current technology the leader can become a monopolist until the time that the new technology arrives.

When the point of time at which the new technology appears on the market lies relatively far in the future, it is optimal for both firms to apply the Compulsive strategy, i.e. buy the current technology immediately and upgrade at time T. This explains why for both firms the Compulsive strategy is optimal when the arrival date of the second technology lies somewhere in the interval $(T_{LeC,C}, \infty)$.

ENDOGENOUS FIRM ROLES

Since firms are identical there seems to be no reason why one of these firms should be given the leader role beforehand. This makes the outcome of the exogenous firm roles case hard to accept. However, here we use this outcome to generate the equilibria of the case where it is not known beforehand which firm will invest first. The fact that firms are identical and rational also implies that it is reasonable to impose that firms behave in the same manner, since no reason of why they should act differently can be given. Therefore we restrict ourselves to symmetric strategies.

Denote the value of the leader and the follower as function of T, the arrival time of the second technology, by $L(T)$ and $F(T)$, respectively. When both firms apply the leader's strategy, their payoff equals $M(T)$. From Table 5.2 and equations (5.9) and (5.11) it follows that

$$L(T) = \begin{cases} e^{-rT}(P_{22} - C_l) & \text{if } T \in [0, T_{LeC,Le}], \\ \begin{aligned} &P_{10} - C_e \\ &+e^{-rT}(P_{22} - P_{10} - C_u) \end{aligned} & \text{if } T \in (T_{LeC,Le}, T_{LeC,C}], \\ \begin{aligned} &P_{11} - C_e \\ &+e^{-rT}(P_{22} - P_{11} - C_u) \end{aligned} & \text{if } T \in (T_{LeC,C}, \infty), \end{cases}$$

(5.17)

$$F(T) = \begin{cases} e^{-rT}(P_{22} - C_l) & \text{if } T \in [0, T_{LeC,C}], \\ \begin{aligned} &P_{11} - C_e \\ &+e^{-rT}(P_{22} - P_{11} - C_u) \end{aligned} & \text{if } T \in (T_{LeC,C}, \infty), \end{cases}$$

(5.18)

$$M(T) = \begin{cases} e^{-rT}(P_{22} - C_l) & \text{if } T \in [0, T_{LeC,Le}], \\ \begin{aligned} &P_{11} - C_e \\ &+e^{-rT}(P_{22} - P_{11} - C_u) \end{aligned} & \text{if } T \in (T_{LeC,Le}, \infty). \end{cases}$$

(5.19)

In Figure 5.1 the payoffs are plotted. This figure should be read as a feedback diagram. Note that contrary to the payoff figures in Chapter 4, the figure shows the payoffs of both firms as function of the arrival time of the second technology T instead of the payoffs as function of time t.

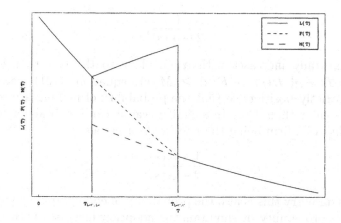

Figure 5.1. Payoffs in scenario 9 as function of T, the arrival time of the second technology, when the firm roles are endogenously determined.

Apart from the interval $(T_{LeC,Le}, T_{LeC,C}]$ the firms choose for the same strategy in the case of exogenous firm roles (see Figure 5.1), meaning that they invest at the same time in the same technology. Hence, there is no difference in the behavior of leader and follower so that we end up with the same equilibria as in the case of endogenous firm roles.

So, what is left to do is to determine the equilibria on the interval $(T_{LeC,Le}, T_{LeC,C}]$. From Figure 5.1 we obtain that the firm that invests first gets the highest payoff, since for $T \in (T_{LeC,Le}, T_{LeC,C}]$ it holds that $L(T) > F(T)$. In case both firms invest at the same time, they both get $M(T)$. This game is a preemption game. In Appendix A.2 of Chapter 4 it was shown that the equilibrium strategy for each firm is given by the following pair of functions

$$G(s) = 1, \tag{5.20}$$

$$\alpha(s) = \frac{L(s) - F(s)}{L(s) - M(s)}, \tag{5.21}$$

where $s \in (T_{LeC,Le}, T_{LeC,C}]$. Recall that $G(s)$ is the cumulative probability that a firm invests at time s given that there was no investment before time s and that $\alpha(s)$ measures the intensity of that probability of investment (cf. Appendix A.2 of Chapter 4).

Of course, both firms do not want to invest at the same time, because it leaves them with the lowest possible payoff M. In Appendix A.2 of Chapter 4 we derived that the probability of occurrence of such a mistake is

$$\frac{\alpha(s)}{2 - \alpha(s)}, \tag{5.22}$$

which naturally increases with $\alpha(s)$. Due to the fact that for $s \in (T_{LeC,Le}, T_{LeC,C}]$ $L(s) > F(s) \geq M(s)$, equation (5.21) learns that $\alpha(s)$ is strictly positive, so that the probability of making a mistake is strictly greater than zero. In a similar way it can be obtained that the probability of a firm being the first investor equals

$$\frac{1 - \alpha(s)}{2 - \alpha(s)}. \tag{5.23}$$

Due to symmetry this is also the probability of ending up being follower. Since the probability of simultaneous adoption increases with $\alpha(s)$, it follows that the probability of being the first investor decreases with $\alpha(s)$, which is at first sight a strange result. But it is not that strange, because if one firm increases its probability to invest, the other firm does the same. This results in a higher probability of making a mistake, which leaves less room for the equal probabilities of being the first investor.

Substitution of the Compulsive and the Leapfrog payoffs, which are given by (5.9) and (5.11), respectively, into (5.21) results in the following expression for $\alpha(s)$ for $s \in (T_{LeC,Le}, T_{LeC,C}]$:

$$\alpha(s) = \frac{P_{10} - C_e + e^{-rs}(-P_{10} + C_l - C_u)}{(P_{10} - P_{11})(1 - e^{-rs})}. \tag{5.24}$$

To compare different situations concerning the arrival time of the new technology, differentiate (5.24) with respect to s, and eventually obtain

$$\frac{\partial \alpha(s)}{\partial s} = \frac{re^{-rs}(C_e - C_l + C_u)}{(P_{10} - P_{11})(1 - e^{-rs})^2} > 0, \tag{5.25}$$

where the inequality sign follows from (5.4). The implication is that if we consider two games, one where the second technology arrives at time T_1 and the other one where the second technology arrives at time T_2, where T_1 and T_2 are related such that $T_{LeC,Le} < T_1 < T_2 < T_{LeC,C}$, then the probability of making the simultaneous adoption mistake in the game where the second technology arrives at time T_1 is smaller than in the game where the second technology arrives at time T_2. To understand this result, consider Figure 5.1: (1) the difference between the payoff from being the leader and the payoff of the follower increases, so that the relative profitability of winning the investment race rises, and (2) the difference between the follower payoff and the payoff that results from simultaneous investment decreases so that firms more and more prefer to win the investment race rather than to make the joint adoption mistake.

Furthermore from (5.24) and the fact that the probability of the joint adoption mistake increases with $\alpha(s)$, the following ceteris paribus results can be derived: the joint adoption mistake is more likely to occur for lower values of C_e or C_u, or for higher values of P_{10}, P_{11}, C_l, or r.

3.4   EQUILIBRIUM STRATEGIES IF LEARNING IS NEGLIGIBLE AND THE NEW TECHNOLOGY IS EXPENSIVE

In this subsection scenario 2 of Table 5.1 is analyzed. From that table we obtain that in this scenario it holds that

$$P_{21} - C_u - P_{11} \leq 0, \tag{5.26}$$

$$P_{21} - C_l > P_{11} - C_d, \tag{5.27}$$

$$P_{22} - C_l \leq P_{12} - C_d. \tag{5.28}$$

In Table 5.1 we see that the cost of upgrading is large in this scenario. This means that, even in case the firm is already active on this market by producing with the current technology, the learning effect is that low that it is still costly to buy and implement the new technology. From (5.26) it can be concluded that, given that the other firm produces with the current technology, it is not profitable to upgrade so that the Compulsive strategy will not be optimal. The same holds when the other firm produces with the new technology, because then exchanging the current technology for the new one is even less profitable.

Taking into account the payoffs of the Leapfrog and the Laggard strategy (cf. (5.11) and (5.12)), we can derive from (5.27) and (5.28) that the Leapfrog strategy is more profitable than the Laggard strategy if the other firm produces with the current technology, while it is the other way round when the other firm produces with the new technology. Since we already concluded that upgrading is never optimal, it follows that in this scenario demand on the output market is too small for two firms producing with the more expensive new technology.

EXOGENOUS FIRM ROLES

Again we first consider the case where one firm is the leader and the other firm the follower. The equilibrium strategies are presented in Table 5.3. Time $T_{B,Le}$ is defined such that at that time the payoffs of Buy and Hold and Leapfrog are equal. The reason for choosing the notation $T_{B,Le}$ here is that any other notation does not fit with our notation introduced earlier. The reason is that the other firm does not play a fixed strategy. At $T_{B,Le}$ it holds that payoffs are equal in case the firm plays Leapfrog while the other plays Buy and Hold, and when the firm plays Buy and Hold and the other firm Leapfrogs. With equations (5.10), (5.11), and (5.12) we derive that

$$T_{LaB,Le} = \frac{1}{r} \log \left(\frac{P_{10} - C_d}{P_{10} - C_e} \right), \qquad (5.29)$$

$$T_{B,Le} = \frac{1}{r} \log \left(\frac{P_{21} - C_l - P_{12} + P_{10}}{P_{10} - C_e} \right), \qquad (5.30)$$

$$T_{LeB,B} = \frac{1}{r} \log \left(\frac{P_{21} - C_l}{P_{11} - C_e} \right). \qquad (5.31)$$

From equations (5.5) and (5.27) it follows that

$$0 < T_{LaB,Le} < T_{B,Le}. \qquad (5.32)$$

Further, we derived that

$$T_{B,Le} < T_{LeB,B}, \qquad (5.33)$$

if and only if the following equation holds,

$$\frac{P_{11} - C_e}{P_{21} - C_l} < \frac{P_{10} - P_{11}}{P_{10} - P_{12}}. \tag{5.34}$$

The economic interpretation of this inequality becomes clear later in this section when we consider the case where $T_{LaB,Le} < T < T_{LeB,B}$.

Table 5.3. Equilibrium strategies in scenario 2 as function of T, the arrival time of the second technology, when the firm roles are assigned exogenously.

T interval	Leader	Follower
$[0, T_{LaB,Le}]$	Leapfrog	Laggard
$(T_{LaB,Le}, T_{B,Le}]$	Leapfrog	Laggard
$(T_{B,Le}, T_{LeB,B}]$	Buy and Hold	Leapfrog
$(T_{LeB,B}, \infty)$	Buy and Hold	Buy and Hold

Since in this scenario the cost of upgrading is too high for a replacement of the current technology by the new technology to be profitable, the Compulsive strategy will never be applied. On the interval $[0, T_{LaB,Le}]$ the arrival time of the new technology is that close that for both firms it is not optimal to invest immediately. One firm will adopt the new technology as soon as it arrives. The other firm waits until time T to acquire the current technology, since from this time onwards the acquisition cost of the current technology is lower (cf. (5.5)). Note that, given the fact that one firm plays Leapfrog, expression (5.28) implies that for the other firm the Laggard strategy is most profitable. The reason is that the demand for output is too low for the two firms to produce both with the new expensive technology.

On the time interval $(T_{LaB,Le}, T_{LeB,B}]$ T is a bit further away, which implies that, given that the leader follows a Buy and Hold strategy, the follower will prefer Leapfrog. The earnings that arise from producing with the current technology on the time interval before time T are not large enough for the follower to justify the immediate acquisition of the current technology. Note that, given that the leader applies a Buy and Hold strategy, for the follower the Leapfrog payoff is higher than the Laggard payoff (cf. (5.27)). The fact that the follower plays Leapfrog implies that until the new technology arrives the leader is the only producer on the market. This monopoly position increases revenue before time T, compared to the situation where both firms apply Buy and Hold. On the other hand, after time T the Buy and Hold strategy will generate less revenue, because then the other firm captures a larger share of the market since it produces more efficiently with the new technology. Hence, two opposite effects are working on the Buy and Hold payoff once the follower switches to Leapfrog. In Table 5.3 and in Figure 5.2 (presented later on) the monopoly effect is assumed to dominate, which explains the downward jump of the Buy and Hold (leader) payoff right at $T_{LeB,B}$. But the reverse can also be true. After straightforward calculations it can be concluded that the jump is indeed downward in case (5.34) holds. We see that in any case at the beginning of the interval $(T_{LaB,Le}, T_{LeB,B}]$ the Buy and Hold payoff is less than the Leapfrog payoff so that the first investor earns less than the follower. Recall that $T_{B,Le}$ is the point of time at which the payoffs of Buy and Hold and Leapfrog are equal (see Figure 5.2, note that $T_{B,Le}$ does not exist in case (5.34) does not hold). Then, on the interval $(T_{LaB,Le}, T_{B,Le}]$ (or $(T_{LaB,Le}, T_{LeB,B}]$ if (5.34) does not hold) the leader refrains from investment and applies a Leapfrog strategy. The follower is not allowed to invest earlier than the leader, so he has to choose between Leapfrog and Laggard. The follower's choice will be Laggard, since this leaves him with the highest payoff (cf. (5.34)). In case equation (5.34) holds we have to consider the interval $(T_{B,Le}, T_{LeB,B}]$. For the leader it is optimal to apply the Buy and Hold strategy and the follower responds by using the Leapfrog strategy.

When the point of time T at which the new technology appears on the market is relatively far away, it is optimal for both firms to start producing with the current technology. Therefore, both firms apply the Buy and Hold strategy if the arrival time of the second technology belongs to the interval $(T_{LeB,B}, \infty)$.

ENDOGENOUS FIRM ROLES

Here it is not specified beforehand which firm will be the first investor, so that both firms can be leader or follower. As said before this seems the proper way to analyze a duopoly with identical firms. Since firms are identical, we restrict ourselves to symmetric strategies. From Table 5.3 and equations (5.10), (5.11), and (5.12) it follows that (with one modification which will be explained right after (5.37))

$$
L\left(T\right) = \begin{cases} e^{-rT}\left(P_{21} - C_l\right) & \text{if } T \in [0, T_{LaB,Le}], \\ P_{10} - C_e + e^{-rT}\left(P_{12} - P_{10}\right) & \text{if } T \in (T_{LaB,Le}, T_{LeB,B}], \\ P_{11} - C_e & \text{if } T \in (T_{LeB,B}, \infty), \end{cases}
$$

$$(5.35)$$

$$F\left(T\right) = \begin{cases} e^{-rT}\left(P_{12} - C_d\right) & \text{if } T \in [0, T_{LaB,Le}], \\ e^{-rT}\left(P_{21} - C_l\right) & \text{if } T \in (T_{LaB,Le}, T_{LeB,B}], \\ P_{11} - C_e & \text{if } T \in (T_{LeB,B}, \infty), \end{cases} \qquad (5.36)$$

$$M\left(T\right) = \begin{cases} e^{-rT}\left(P_{22} - C_l\right) & \text{if } T \in [0, T_{LaB,Le}], \\ P_{11} - C_e & \text{if } T \in (T_{LaB,Le}, \infty). \end{cases} \qquad (5.37)$$

Note that now the leader plays the Buy and Hold strategy in the games where the second technology arrives somewhere in the interval $(T_{LaB,Le}, T_{B,Le}]$, because a firm can only become leader by investing as first. The expected value of waiting equals the Laggard payoff which is less than the payoff resulting from the Buy and Hold strategy. Therefore one firm will invest in the current technology. In Figure 5.2 the payoffs are plotted.

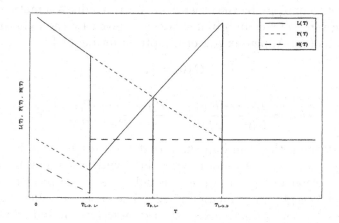

Figure 5.2. Payoffs in scenario 2 as function of T, the arrival time of the second technology, when the firm roles are endogenously determined.

From Figure 5.2 it is obtained that only for the interval $(T_{LeB,B}, \infty)$ the strategies of both firms are identical, implying that the equilibrium for games where $T \in (T_{LeB,B}, \infty)$ is the same for the exogenous and the endogenous firm roles case.

To solve the whole problem we divide the remaining time interval into three subintervals and treat the three cases in order of timing. For the games with arrival date of the second technology in the interval $[0, T_{LaB,Le}]$ the optimal strategies are Leapfrog and Laggard, i.e. no investment takes place before time T. The firms wait until the arrival of the new technology after which the following game will be played.

Without loss of generality we set $T = 0$ for the moment. Since the highest payoff can be obtained by a Leapfrog strategy (provided that the other firm plays Laggard), it is attractive to be the first investor in the new technology. However, when both firms apply the strategy invest right away in the new technology at time T, with probability one they end up with the payoff $M(0)$, which is less than what could be obtained by investing in the current technology, $F(0)$. Hence, it seems that a mixed strategy is called for in this preemption game. It is important to realize that we only need to take into account positive probabilities for investment in the new technology. The reason for considering an investment in the current technology would be that the firm does not want to end up in a situation where they both have invested in the new technology. However, the probability of reaching this situation only depends on the probability of investing in the new technology. Hence, investing in the current technology will only occur after an investment in the new technology by the other firm. In Appendix A a mathematical proof is given. The equilibrium strategy is given by the following pair of functions (see Appendix A.2 of Chapter 4 for details)

$$G(0) = 1, \tag{5.38}$$

$$\alpha(0) = \frac{L(0) - F(0)}{L(0) - M(0)} = \frac{P_{21} - C_l - (P_{12} - C_d)}{P_{21} - P_{22}}. \tag{5.39}$$

The worst thing that can happen is that both firms invest in the new technology at the same time, leaving them with a low payoff $M(0) = P_{22} - C_l$. Analogous to the previous subsection, the probability that this happens is given by equation (5.22). The expected value of the firm resulting from the optimal mixed strategy described by equations (5.38) and (5.39) equals the payoff associated with the Laggard strategy (see Appendix A.2 of Chapter 4 where it was shown that the expected payoff in a preemption game equals the follower payoff). Therefore, the value of the firm is also equal to the Laggard payoff if $T \in (0, T_{LaB,Le}]$, and from Figure 5.2 it can be concluded that this value decreases in T.

It turns out to be convenient to divide the interval $(T_{LaB,Le}, T_{LeB,B}]$ into two parts: $(T_{LaB,Le}, T_{B,Le}]$ and $(T_{B,Le}, T_{LeB,B}]$. Note that $T_{B,Le} \in (T_{LaB,Le}, T_{LeB,B}]$ only in case (5.34) holds. If (5.34) does not hold, then during the whole interval $(T_{LaB,Le}, T_{LeB,B}]$ the Buy and Hold payoff falls below the Leapfrog payoff. Let us first consider $(T_{LaB,Le}, T_{B,Le}]$. Here the new technology will become available soon, which implies that the Leapfrog strategy is more profitable than Buy and Hold. Therefore, the payoff of the first investor is lower than the payoff of the follower, so that a second mover advantage arises. Hence, each firm prefers to be the follower, but if it has to be the first investor, it prefers to invest earlier rather than later, because the Buy and Hold payoff falls over time. Thus the game is an attrition game. A possible strategy would be to refrain from investment during this interval, waiting for the other firm to invest. Since with identical firms there is no reason to believe why the other firm would act differently, nothing happens on this interval. This implies that both firms end up with playing the game at time T, which is described above. The expected value of the firm obtained from playing this game equals the Laggard payoff, and this payoff lies below the payoffs of both Buy and Hold and Leapfrog. In Appendix A.3 of Chapter 4 the subgame perfect equilibrium is given for an attrition game with identical players, as it occurs here. The equilibrium strategy is described by (note that due to the difference in definition of the payoff functions L, F, and M in this chapter the argument of the functions is $T - v$ instead of v)

$$G(s) = 1 - e^{\int_{v=0}^{s} \frac{dL(T-v)}{F(T-v)-L(T-v)}}, \tag{5.40}$$

$$\alpha(s) = 0, \tag{5.41}$$

where $s \in [0, T - T_{LaB,Le})$. Equation (5.41) implies that the probability that the two firms move exactly at the same time is zero, which explains why M does not affect the mixed investment strategy in (5.40).

From (5.40) it is easily obtained that for $T \in (T_{LaB,Le}, T_{B,Le}]$ it holds that

$$\lim_{s \uparrow T - T_{LaB,Le}} G(s) < 1.$$

Note that, if this were not the case, the symmetric investment strategy is not a Nash equilibrium. The reason is that when

$$\lim_{s \uparrow T - T_{LaB,Le}} G(s) = 1,$$

then one of the firms could do better by refraining from investment during the interval $[0, T - T_{LaB,Le})$, since this firm knows for sure that its competitor will have invested at some time before $T - T_{LaB,Le}$.

Substitution of the relevant formulas for the payoffs into (5.40) leads to

$$G(s) = 1 - e^{\int_{v=0}^{s} \frac{re^{-r(T-v)}(P_{12}-P_{10})}{C_e - P_{10} + e^{-r(T-v)}(P_{21}-C_l-P_{12}+P_{10})} dv}. \tag{5.42}$$

The firm's willingness to invest increases with the relative performance of Buy and Hold compared to Leapfrog. In this light the following ceteris paribus results, that are derived from (5.42), are easy to understand: G goes up with r, T, P_{10}, C_l, and goes down with C_e and P_{21}.

Finally, we analyze games for which $T \in (T_{B,Le}, T_{LeB,B}]$. Solving this case leads to analogous results as in the previous subsection. In the symmetric equilibrium both firms use the following strategy

$$G(s) = 1, \tag{5.43}$$

$$\alpha(s) = \frac{L(s) - F(s)}{L(s) - M(s)}, \tag{5.44}$$

where $s \in (T_{B,Le}, T_{LeB,B}]$. Right at the start of the game one of the firms will invest in the current technology. The other firm refrains from investment until time T at which it will adopt the new technology. The probability that both firms invest in the current technology exactly at the same time is again given by equation (5.22). It is unclear how $\alpha(s)$ develops over time, since two opposite effects are working here: (1) the difference in payoffs between leading and following decreases over time which has a negative effect on $\alpha(s)$, and (2) the difference in payoffs between leading and joint adoption decreases as time passes which has a positive effect on $\alpha(s)$.

4.   CONCLUSIONS

This chapter treats the technology adoption decision of the firm in a duopoly framework. One of the main difficulties concerning the technology investment decision in practice is that in the future better technologies than now available will be invented. The model in this chapter tries to capture this important aspect by considering two technologies: one which is available immediately, and the other one which is more efficient and becomes available at a known point of time in the future. By doing so, work of Reinganum, 1981, Fudenberg and Tirole, 1985, Hendricks, 1992, and Stenbacka and Tombak, 1994, who consider only one technology, is extended. Moreover, learning is incorporated in the way that adoption of the current technology makes it less costly to adopt and implement the new technology.

We focussed on the scenario where for every technology investment it holds that the discounted future profit stream exceeds the immediate expenditure. In case the arrival date of the new technology lies far in the future, the future presence of a new technology does not prevent that investing in the current technology is still optimal. When this date comes nearer it is not optimal anymore for both firms to invest in the current technology right away. Hence, one of the firms has to refrain from investment, which implies that by investing in the current technology the other firm obtains monopoly profits until the arrival date of the new technology. To capture these monopoly profits a firm must try to invest earlier than its competitor. In this way the preemption equilibria arise that we already know from, e.g., Fudenberg and Tirole, 1985, but here no rent equalization occurs as was the case in that paper. A consequence of the absence of rent equalization is that a positive probability arises that both firms invest at the same time, leaving them with a very low payoff (in Fudenberg and Tirole, 1985 the probability of occurrence of this mistake was zero due to rent equalization).

Another new element in our chapter is the occurrence of second mover advantages in technology adoption problems. This happens in scenarios where technology upgrading is not optimal so that firms have to make a choice between investing in the current technology right away and keep on producing with it, or waiting with investment until the new technology arrives. The advantage of the immediate investor is that monopoly profits are gained until the arrival of the new technology, while the investor in the new technology has the advantage of producing with a better technology once it is available. A second mover advantage arises when the advantage of producing with the new technology in the future leads to a higher payoff than the current temporary monopoly profits.

An immediate extension of the model in this chapter is to add uncertainty. A distinction can be made between uncertainty concerning the arrival date of new technologies or uncertainty concerning the efficiency of new technologies. For instance, in case of micro-chips the technical parameters and specifications of future designs are known beforehand, but the arrival date is uncertain since the appearance of the technology depends on research and development and the market factors affecting the introduction of the product (see Chapter 6 and Nair, 1995).

Another interesting topic of future research is to incorporate asymmetric information in the sense that a firm does not know how profitable a particular innovation is for its opponent. However, as time passes the firm learns about the other firm's profit function from the observed investment behavior of the other firm. Based on this observation it will update its conjecture about the other firm's profit function (see Lambrecht and Perraudin, 1999).

Appendix

A.   PROOF

Define α_i (β_i) to be the probability that firm 1(2) invests in technology i ($i = 1, 2$). Let us consider the investment decision of firm 1. Concerning the payoffs we introduce the following notation: L is the value of firm 1 if firm 1 invests in the new technology (= technology 2) and firm 2 invests in the current technology (=technology 1), F is the value of firm 1 if firm 1 invests in the current technology, and firm 2 invests in the new technology. M_1 is the value of firm 1 if both firms invest in the current technology, and M_2 is the value of firm 1 if both firms invest in the new technology. It is easily verified that under the conditions given in Section 3.4 it holds that

$$L > M_1 > F > M_2.$$

The value of firm 1 under the strategies (α_1, α_2) and (β_1, β_2) equals

$$V(\alpha_1, \alpha_2, \beta_1, \beta_2) = \frac{\alpha_2 \beta_2 M_2 + \alpha_2 \beta_1 L + \alpha_2 (1 - \beta_1 - \beta_2) L}{1 - (1 - \alpha_1 - \alpha_2)(1 - \beta_1 - \beta_2)}$$
$$+ \frac{\alpha_1 \beta_2 F + \alpha_1 \beta_1 M_1 + \alpha_1 (1 - \beta_1 - \beta_2) F}{1 - (1 - \alpha_1 - \alpha_2)(1 - \beta_1 - \beta_2)}$$
$$+ \frac{(1 - \alpha_1 - \alpha_2) \beta_2 F + (1 - \alpha_1 - \alpha_2) \beta_1 L}{1 - (1 - \alpha_1 - \alpha_2)(1 - \beta_1 - \beta_2)}.$$

After deriving the first order conditions for optimality and considering symmetric strategies it turns out that there is a unique symmetric equilibrium given by:

$$\alpha_1 = \beta_1 = 0,$$
$$\alpha_2 = \beta_2 = \frac{L - F}{L - M_2}.$$

Chapter 6

MULTIPLE NEW TECHNOLOGIES

1. INTRODUCTION

In this chapter we extend the models of Chapters 4 and 5 by adding uncertainty to the innovation process and by considering multiple new technologies. The new technologies are invented at previously unknown points of time. A comparable framework is considered in the duopoly model by Gaimon, 1989. The difference is that in that paper a continuous stream of new technologies arrives over time, which is known beforehand by the firms.

The investment decision problem of this chapter is solved by introducing the waiting curve as a new concept in timing games. The waiting curve is equal to the expected equilibrium payoff of the firm when both firms wait with making an investment (at least) until the next technology has arrived. Therefore the waiting curve resembles the option to invest in some future technology that is not invented yet.

The remainder of this chapter is organized as follows. In Section 2 the investment decision problem of the firm is described. We reformulate the investment decision problem as a timing game, and design an algorithm to solve it in Section 3. In Section 4 we apply the algorithm to an information technology investment problem. Concluding remarks are given in Section 5.

2. THE MODEL

In this section we describe the model of this chapter. A duopoly is considered where both firms maximize their value over an infinite planning horizon. We define $T\ (\geq 0)$ to be the time elapsed since the start of the game. The first assumption is that firms are identical. Each firm has a profit function $\pi\,(\theta_x, \theta_y)$, where $\theta_x\,(\geq 0)$ equals the technology-efficiency parameter of the technology that the firm uses itself and $\theta_y\,(\geq 0)$ that of its opponent. The profit function of each firm is non-negative, increasing and concave in its own technology-efficiency parameter and decreasing in its rival technology-efficiency parameter. This for the reasons that (i) a firm can make more profits when it produces with a more efficient technology, (ii) the growth of the profits will be limited (due to output market saturation and the fact that production costs are always positive), and (iii) a firm will make less profits when its rival uses a more efficient technology. Furthermore it holds that the impact of the technological improvement of the home firm on its profits is decreasing in the technology efficiency parameter of the other firm. Therefore, the profit function satisfies the following inequalities:

$$\pi\,(\theta_x, \theta_y) \geq 0, \tag{6.1}$$

$$\frac{\partial \pi\,(\theta_x, \theta_y)}{\partial \theta_x} > 0, \tag{6.2}$$

$$\frac{\partial \pi\,(\theta_x, \theta_y)}{\partial \theta_y} < 0, \tag{6.3}$$

$$\frac{\partial^2 \pi\,(\theta_x, \theta_y)}{\partial \theta_x^2} < 0, \tag{6.4}$$

$$\frac{\partial^2 \pi\,(\theta_x, \theta_y)}{\partial \theta_x \partial \theta_y} < 0. \tag{6.5}$$

We analyze a dynamic model with an infinite planning horizon. Risk-neutral firms are considered, which discount the stream of future profits at a constant rate $r\,(> 0)$. Initially, at time $T = 0$ each firm produces with a technology of which the efficiency is designated by $\theta_0\,(\geq 0)$. As time passes new technologies become available at discrete points of time. Technologies become more and more efficient over time, and the more efficient a technology the larger the associated parameter θ. The i-th technology has an efficiency represented by $\theta_i\,(> \theta_{i-1})$, for $i \in \mathbf{N}$. We define $T_i\,(\geq 0)$ to be equal to the point in time at which technology i becomes available, $i \in \mathbf{N}$, and $T_0 = 0$. Each firm has the opportunity to adopt at time $T\,(\geq 0)$ one of the technologies being available at time T by investing $I\,(t_i)$ to adopt technology i, where $t_i\,(\geq 0)$ is the length of the time period passed since the introduction of technology i, i.e.

$t_i = T - T_i$. We assume the second hand market for these capital goods to be negligible (e.g. information technology products) so that this investment is irreversible. The differences between the technologies are all captured in the different values for the efficiency parameter θ, so that, without loss of generality, investment expenditures $(= I(\cdot))$ can be set equal for all technologies. The investment cost $I(\cdot)$ is non-negative, decreasing and convex in time:

$$I(t) > 0, \tag{6.6}$$

$$\frac{\partial I(t)}{\partial t} < 0, \tag{6.7}$$

$$\frac{\partial^2 I(t)}{\partial t^2} > 0. \tag{6.8}$$

Such a decrease can be motivated by the fact that better technologies become available as time passes so that the demand for the current technology decreases over time. Another factor can be learning by doing in the production process of the technology supplier.

Furthermore, we assume, as anywhere else, that the process of technological evolution (innovation supply) is exogenous to the firms. The arrival process of the new technologies is a stochastic process. We assume that the associated increases in θ are known beforehand. In practice this occurs, for example, in the case of micro-chips where the technical parameters and specifications of future designs are known beforehand, but the arrival date is uncertain since the appearance of technology depends on research and development and market factors affecting the introduction of the product (see also Nair, 1995).

At time T the number $N(T)$ refers to the technology that became last available. To incorporate the uncertainty in the innovation process we assume that $N(T)$ is a Poisson process with rate $\lambda (> 0)$. The interarrival time $\tau_i (\geq 0)$ is the time between the invention times of the $(i-1)$-th and i-th technology: $\tau_i = T_i - T_{i-1}$, $i \in \mathbf{N}$. As a result of the Poisson arrival process the τ_i's are independently and identically distributed according to an exponential distribution with parameter λ.

3.   TIMING GAME

For simplicity reasons we restrict ourselves to the case where firms can only make one technology switch. This typically holds for firms whose financial means are limited. We transform the investment decision problem into a two player timing game. In Appendix A of Chapter

4 a rigorous treatment of timing games is given. Here we repeat the important features.

In a timing game each player has to decide when to make a single move. The player that moves first is called the leader and the other is the follower. Since firms are identical there seems to be no reason why one of these firms should be given the leader role beforehand. Therefore, we strive at obtaining equilibria where it is not known beforehand which firm will invest first. In the general setting of a timing game the payoff of a player depends on its own date of moving and the other player's date of moving. In case one player has already moved, the problem for the other player is a one person decision problem. A player can react instantaneously to its opponent's action.

Four payoff curves are important in our timing game. Each payoff curve is a function of time $t\,(\geq 0)$, which is the time passed since the last technology has become available for the firms: $t\,(T) = T - T_{N(T)}$. In the remainder of this chapter we write t instead of $t\,(T)$ whenever there is no confusion possible. Let the leader move at time t. Then the value of the follower, which is the outcome of the one person's decision problem, is denoted by $F\,(t)$. The value of the leader is given by $L\,(t)$, in which the optimal action of the follower is included. In case of a simultaneous move at time t the value of a player is denoted by $M\,(t)$. Since simultaneous moving is always possible for the follower, it holds that

$$F\,(t) \geq M\,(t), \text{ for } t \geq 0. \qquad (6.9)$$

The fourth curve is called the waiting curve, which is a new concept within the area of timing games. Here, the waiting curve is used to transform the investment decision problem under consideration into a timing game. The waiting curve represents the expected payoff of a firm if both firms do not move (at least) until the next arrival of a new technology and act optimally afterwards. This implies that we need to know the equilibrium outcome of the game that starts after the arrival of a new technology. As a result we have to consider a finite number of new technologies. Due to discounting this assumption is not too strict. In order to find the right number of new technologies to take into account in the model, the following algorithm, which is a weak forecast horizon procedure, can be used:

Step 0 Solve the model with one technology.

Step 1 Add one extra technology to the model and solve the model.

Step 2 If the results of the last two models are very different go to step 1, otherwise the right number of technologies has been found.

A model with n new technologies is solved as follows. Start with solving the timing game that starts after the arrival of the n-th technology. This game is a classical timing game, since it contains no waiting curve. The equilibrium outcomes of this game are used to construct the waiting curve for the game that starts at some time during the interval $[T_{n-1}, T_n)$. Solve this game and use the equilibrium outcomes to construct the waiting curve for the game that starts somewhere at the time interval $[T_{n-2}, T_{n-1})$. This procedure goes on until the game that starts at time $T_1 = 0$ is solved.

This section describes the construction of the four payoff curves. In Subsection 3.1 we derive the value of a firm given each firm's strategy. Using this value function, we determine the leader, follower and joint-moving curves in Subsection 3.2. In Subsection 3.3 possible equilibria of timing games without waiting curve are considered. The waiting curve is constructed in Subsection 3.4. In Subsection 3.5 we explain the implication of adding the waiting curve for the possible equilibria of timing games. Finally, in Subsection 3.6 the algorithm for solving the investment decision problem with a finite number of new technologies is summarized.

3.1 VALUE FUNCTION

In the investment decision problem firms not only have to decide when to adopt a technology, but also which technology to adopt. Define $V(S, i, T, j)$ as the expected value at time T of a firm that adopts technology i at time $S \geq T$ itself, while its rival adopts technology j at time T. Of course, it must be true that $T \geq T_j$ and $S \geq T_i$. The expected value of the firm at time T equals

$$V(S, i, T, j) = E\left[\int_{u=0}^{S-T} \pi(\theta_0, \theta_j) e^{-ru} du \right. \tag{6.10}$$
$$\left. + \int_{u=S-T}^{\infty} \pi(\theta_i, \theta_j) e^{-ru} du - I(S - T_i) e^{-r(S-T)} \right].$$

The expected value of the firm's opponent at time T is equal to

$$V(T, j, S, i) = E\left[\int_{u=0}^{S-T} \pi(\theta_j, \theta_0) e^{-ru} du \right.$$

$$\left. + \int_{u=S-T}^{\infty} \pi(\theta_j, \theta_i) e^{-ru} du - I(T - T_j)\right]. \tag{6.11}$$

Rewriting (6.10) gives

$$V(S, i, T, j) = \frac{\pi(\theta_0, \theta_j)}{r}\left(1 - E\left[e^{-r(S-T)}\right]\right) \tag{6.12}$$

$$+ \frac{\pi(\theta_i, \theta_j)}{r} E\left[e^{-r(S-T)}\right] - I(S - T_i) E\left[e^{-r(S-T)}\right].$$

Equation (6.11) can be written as follows

$$V(T, j, S, i) = \frac{\pi(\theta_j, \theta_0)}{r}\left(1 - E\left[e^{-r(S-T)}\right]\right) \tag{6.13}$$

$$+ \frac{\pi(\theta_j, \theta_i)}{r} E\left[e^{-r(S-T)}\right] - I(T - T_j).$$

For determining the value functions (6.12) and (6.13) there is one thing left to derive: an expression for $E\left[e^{-r(S-T)}\right]$. Recall that $N(T)$ is the index of the most efficient technology that is available at time T. We distinguish two cases: in the first case the second investor wants to invest in an already existing technology, while in the second case this firm plans to invest in a technology that does not exist yet. In the first case, it holds that $N(T) \geq i$, and therefore the value of S is known for sure at time T:

$$E\left[e^{-r(S-T)}\big| N(T) \geq i\right] = e^{-r(S-T)}. \tag{6.14}$$

Now consider the second case where $N(T) < i$, then

$$E\left[e^{-r(S-T)}\big| N(T) < i\right] = e^{-r(S-T_i)} E\left[e^{-r(T_i-T)}\big| N(T) < i\right]. \tag{6.15}$$

Lemma 2.3 states that

$$E\left[e^{-r(T_i-T)}\big| N(T) < i\right] = \left(\frac{\lambda}{r+\lambda}\right)^{i-N(T)}. \tag{6.16}$$

With the help of equation (6.16) we derive that

$$E\left[e^{-r(S-T)}\right] = \begin{cases} e^{-r(S-T)} & \text{if } S < T_{N(T)+1}, \\ e^{-r(S-T_i)}\left(\frac{\lambda}{r+\lambda}\right)^{i-N(T)} & \text{if } S \geq T_{N(T)+1}. \end{cases} \tag{6.17}$$

3.2   LEADER, FOLLOWER AND JOINT-MOVING CURVES

At each point of time T the leader can choose to immediately invest in a technology j from the finite set $\{1, 2, \ldots, N(T)\}$. Given an adoption strategy of the leader (T, j) the optimal reaction of the follower can be calculated in two steps.

In the first step, derive for each technology i the optimal adoption date S_i^* for the follower. Since the follower's payoff depends on the adoption strategy the other firm uses, S_i^* is a function of T and j. Therefore,

$$S_i^*(T, j) = \arg \max_{u \geq \max(T_i, T)} V(u, i, T, j). \tag{6.18}$$

In order to be more specific about $S_i^*(T, j)$ consider the following scenario: the leader has already adopted technology j and technology i has just been invented. The follower can either adopt technology i right away or delay adoption. Let $w_i^F(j)$ denote the optimal waiting time for the follower, that is the length of the time period between invention and optimal adoption of technology i. Solving the maximization problem (6.18) yields that $w_i^F(j) = 0$ if

$$\pi(\theta_i, \theta_j) - \pi(\theta_0, \theta_j) \geq rI(0) - \left. \frac{\partial I(t)}{\partial t} \right|_{t=0}, \tag{6.19}$$

and that $w_i^F(j)$ is implicitly determined by

$$\pi(\theta_i, \theta_j) - \pi(\theta_0, \theta_j) = rI\left(w_i^F(j)\right) - \left. \frac{\partial I(t)}{\partial t} \right|_{t=w_i^F(j)}, \tag{6.20}$$

otherwise. Equation (6.20) states that the marginal costs (the left-hand side) and the marginal benefits (the right-hand side) are equal at time $w_i^F(j)$. The marginal costs are equal to the opportunity costs of the investment, $rI(t)$, and the costs resulting from the fact that the firm invests right away so that it does not take advantage from I being decreasing over time. If at time T_i the marginal benefits exceed the marginal costs (cf. equation (6.19)) the firm should adopt immediately so that $w_i^F(j) = 0$. Using the definition of $w_i^F(j)$ and since $\pi(\theta_i, \theta_j)$ is increasing in θ_i, we know that there exists an $\widehat{i}(j)$ such that $w_i^F(j) = 0$ for all $i \geq \widehat{i}(j)$.

We extend this particular scenario to the general case and conclude that the optimal adoption time $S_i^*(T, j)$ is equal to

$$S_i^*(T, j) = \begin{cases} T & \text{if } T \geq T_i + w_i^F(j), \\ T_i + w_i^F(j) & \text{if } T < T_i + w_i^F(j). \end{cases} \tag{6.21}$$

In the second step, we use (6.21) to determine the technology i^* that maximizes the follower's payoff, given that the leader invests at time T in technology j:

$$i^* (T, j) = \arg \max_k V \left(S_k^* (T, j), k, T, j \right). \qquad (6.22)$$

The leader, on its turn, takes into account the follower's investment behavior in choosing at time T the technology $j^* (T)$ that results in the largest payoff:

$$j^* (T) = \arg \max_{k \in \{1, 2, \dots, N(T)\}} V \left(T, k, S_{i^*(T,k)}^* (T, k), i^* (T, k) \right). \qquad (6.23)$$

The process described above results in the following value functions for the timing game that starts at time $T_k \le T$:

$$L(t) = L(T - T_k) = \left(1 - e^{-r(T - T_k)} \right) \frac{\pi (\theta_0, \theta_0)}{r} \qquad (6.24)$$
$$+ e^{-r(T - T_k)} V \left(T, j^* (T), S_{i^*(T,j^*(T))}^* (T, j^* (T)), i^* (T, j^* (T)) \right),$$

$$F(t) = F(T - T_k) = \left(1 - e^{-r(T - T_k)} \right) \frac{\pi (\theta_0, \theta_0)}{r} \qquad (6.25)$$
$$+ e^{-r(T - T_k)} V \left(S_{i^*(T,j^*(T))}^* (T, j^* (T)), i^* (T, j^* (T)), T, j^* (T) \right),$$

$$M(t) = M(T - T_k) = \left(1 - e^{-r(T - T_k)} \right) \frac{\pi (\theta_0, \theta_0)}{r} \qquad (6.26)$$
$$+ e^{-r(T - T_k)} V (T, j^* (T), T, j^* (T)).$$

3.3   EQUILIBRIA FOR TIMING GAMES WITHOUT WAITING CURVE

In this subsection possible equilibria for classical timing games, i.e. timing games without waiting curves, are presented. In our model with n new technologies, the game that starts after time T_n is a classical timing game.

Classical timing games can be divided in two classes. The first class consists of the so-called preemption games and the elements of the second class are called attrition games. Preemption games are characterized by the fact that there exists a point of time where there is a first mover advantage:

$$\exists t \in [0, \infty) \text{ such that } L(t) > F(t). \qquad (6.27)$$

In an attrition game the follower's payoff exceeds the leader's payoff at all times:

$$F(t) > L(t) \text{ for all } t \in [0, \infty). \qquad (6.28)$$

In general a (classical) timing game can be split up into countably many subgames, where each subgame is a preemption game or an attrition game. Due to the definitions of preemption and attrition games, the split up points will be the points at which the function $L(t) - F(t)$ changes its sign. The equilibrium of a general timing game is found by first solving the last subgame, then using the resulting value functions of the equilibrium of this subgame in the second last subgame and so forth and so on.

Since we analyze identical firms we are especially interested in equilibria with symmetric strategies. For identical and rational firms there is no reason why they should act differently. For a rigorous treatment and a literature overview of classical timing games we refer to Appendix A of Chapter 4.

The equilibrium outcome of the timing game that starts after time T_n depends on the interarrival time τ_n. We denote the (expected) equilibrium outcome of the game that starts after time T_n by $\Omega_n(\tau_n)$. If the game has more than one equilibrium, we use the most reasonable equilibrium in the calculations, being the equilibrium under which the player's payoffs are maximal (the Pareto optimal equilibrium, cf. Appendix A of Chapter 4).

3.4   WAITING CURVE

In general, the equilibrium outcome of the game that starts in the interval $[T_k, T_{k+1})$ is denoted by $\Omega_k(\tau_k)$, with $k \in \{0, \ldots, n-1\}$. Using this notation, the waiting curve for a game that starts in the interval $[T_{k-1}, T_k)$ equals

$$W(t) = W(T - T_{k-1}) = \left(1 - e^{-r(T-T_{k-1})}\right)\frac{\pi(\theta_0, \theta_0)}{r} + e^{-r(T-T_{k-1})}$$

$$\times \int_{\tau_k=0}^{\infty} \left[\int_{u=0}^{\tau_k} \pi(\theta_0, \theta_0) e^{-ru}du + e^{-r\tau_k}\Omega_k(\tau_k)\right]\lambda e^{-\lambda\tau_k}d\tau_k. \qquad (6.29)$$

The first part represents the profits made by the firm on the time interval $[T_{k-1}, T]$. The second part resembles the expected payoff of the firm from time T onwards conditioned on the interarrival time τ_k.

The waiting curve represents the option to invest in some future technology that is not invented yet. As such it is not equal to the option value of waiting since it does not take into account the increased profitability over time (due to the decreasing investment costs) of the already existing technologies.

3.5   EQUILIBRIA FOR TIMING GAMES WITH WAITING CURVE

The equilibria of a timing game with waiting curve are found in two steps. In the first step the timing game is split up into subgames. This is done as is described in Subsection 3.3, i.e. the split up points are the points at which the function $L(t) - F(t)$ changes its sign. In the second step the subgames are solved. The last subgame is solved as first, then the second last subgame, and so forth and so on.

The first class of subgames with waiting curve are those in which the leader curve exceeds the waiting curve for all points in time. The implication is that for the leader investing dominates waiting. Consequently, the equilibria of such a subgame are given by the equilibria of the corresponding subgame without waiting curve.

In a subgame for which the waiting curve exceeds the leader curve, i.e. $W(t) > L(t)$ for all $t \geq 0$, none of the firms is going to invest as first. This for the reason that waiting gives them a higher expected value. Therefore, the equilibrium outcome for both firms is waiting.

If in a subgame the leader curve exceeds the waiting curve for some but not all points in time, then for at least one firm investing is better than waiting for those points in time. There are two cases: (i) the subgame without the waiting curve is a preemption game, and (ii) the subgame without the waiting curve is an attrition game.

In the first case (preemption game) the equilibria of the subgame with waiting curve are given by the equilibria of the subgame without waiting curve. This is a direct result of the fact that the equilibria in the subgame without waiting curve are Nash equilibria, i.e. one firm can not improve his expected value by deviating from the equilibrium strategy.

Contrary, in the second case (attrition game) the firm that is leader can increase its profit by waiting with investing at the points in time where the waiting curve exceeds the leader curve. The equilibrium strategies for the part of the subgame where the leader curve exceeds the waiting curve are given by the equilibrium strategies of the subgame without waiting curve. At the points in time of the other part of the subgame none of the firms will invest.

3.6   SOLUTION PROCEDURE

In this subsection the solution procedure is summarized. In the first step of the solution procedure the classical timing game that starts at time T_n is solved. This gives the equilibrium outcome function $\Omega_n(\tau_n)$. Using this equilibrium outcome function we construct the waiting curve (6.29) and solve the timing game that starts at a point in time on the interval $[T_{n-1}, T_n)$. The resulting outcomes are incorporated in the function $\Omega_{n-1}(\tau_{n-1})$ which is again used to construct the waiting curve for the timing game that starts somewhere during the time interval $[T_{n-2}, T_{n-1})$. This process is repeated until the game that starts at T_1 is solved.

Combining the equilibrium strategies of each step gives the optimal investment strategy of the firm. The ex-ante probabilities of each equilibrium outcome can be derived using the calculations of each step. After each realization of an interarrival time these probabilities must be updated.

4.   INFORMATION TECHNOLOGY INVESTMENT PROBLEM

In this section we apply the algorithm of the previous section to a specific information technology investment problem. Information technology products are heavily dependent on micro-chips. The memory and arithmetic power of micro-chips develop in an exponential way over time. This was firstly recognized by Gordon Moore, one of the Intel-founders, in 1964, who found out that the amount of information on a piece of silicium doubles every year. This statement is called Moore's law. Nowadays, Moore's law still applies although the doubling time has risen to two to three years. In our calculations it is assumed that on average every three years a new generation of chips arrives: $\lambda = \frac{1}{3}$. A new generation of chips is a generation that is twice as efficient as the preceding generation. After applying the algorithm stated in the beginning of Section 3, it turned out that we need to take four generations of chips into account. When we normalize the technology parameter of the current technology to one, this gives rise to the following scheme,

$$\theta_0 = 1, \tag{6.30}$$
$$\theta_{i+1} = 2\theta_i, \ i \in \{0, 1, 2, 3\}, \tag{6.31}$$

so that

$$\theta_i = 2^i, \ i \in \{0, 1, 2, 3, 4\}. \tag{6.32}$$

Due to the rapid innovation process, prices of information technology products go down quickly. We assume that

$$I(t) = I_0 e^{-\alpha t}, \tag{6.33}$$

where

$$I_0 = 50, \tag{6.34}$$
$$\alpha = 1. \tag{6.35}$$

The strange thing with micro-electronics is that their fast efficiency improvement does not impress consumers. As an illustration, consider a telephone in which a certain amount of telephone numbers can be stored. A new generation of chips doubles this amount, but most likely this will not be a reason for customers to sell their old telephone and buy a new one. Another example is that a new generation of personal computers will not double the research output of a scientist. Therefore, a manager of Philips, Claassen (see Rozendaal, 1998), has argued that utility is a logarithmic function of technology, in the sense that utility increases with one unit in case technology power becomes ten times as large. For this reason we assume that profit increases with the technology-efficiency parameter in a logarithmic way with base 10 (cf. (6.32)):

$$\pi\left(\theta_i, \theta_j\right) = \frac{{}^{10}\log\left(2\theta_i^2\right)}{{}^{10}\log\left(2\theta_j\right)} = \frac{2i+1}{j+1}. \tag{6.36}$$

The discount rate equals $r = 0.05$. From equations (6.19), (6.20), (6.33), and (6.36) we derive that

$$w_i^F\left(j\right) = \begin{cases} \frac{1}{\alpha}\log\left(\frac{(r+\alpha)I_0(j+1)}{2i}\right) & \text{if } i < \frac{1}{2}\left(j+1\right)\left(r+\alpha\right)I_0, \\ 0 & \text{else.} \end{cases} \tag{6.37}$$

In Appendix A the expected equilibrium outcomes for the subgames starting right at the invention times T_4, T_3, and T_2 are derived. The results are summarized in Tables 6.1-6.3. In the tables the following leader adoption times are used

$$
\begin{aligned}
t_4^P &= \min\left(t\left|V\left(T_4 + t, 4, T_4 + w_4^F\left(4\right), 4\right)\right.\right) \\
&= V\left(T_4 + w_4^F\left(4\right), 4, T_4 + t, 4\right) = 0.734579,
\end{aligned}
$$

$$
\begin{aligned}
t_{34}^L\left(\tau_4\right) &= \min\left(t\left|V\left(T_4 + t, 3, T_4 + w_4^F\left(3\right), 4\right)\right.\right) \\
&= V\left(T_4 + t, 4, T_4 + w_4^F\left(4\right), 4\right),
\end{aligned}
$$

$$
S_{34}^L\left(\tau_4\right) = \arg\max_{t\in\left[0, t_{34}^L\left(\tau_4\right)\right]} V\left(T_4 + t, 3, T_4 + w_4^F\left(3\right), 4\right),
$$

$$
\begin{aligned}
t_{34}^P\left(\tau_4\right) &= \min\left(t\left|V\left(T_4 + t, 3, T_4 + w_4^F\left(3\right), 4\right)\right.\right) \\
&= V\left(T_4 + w_4^F\left(3\right), 4, T_4 + t, 3\right),
\end{aligned}
$$

$$
\begin{aligned}
t_{34}^P &= \min\left(t\left|V\left(T_3 + t, 3, T_4 + w_4^F\left(3\right), 4\right)\right.\right) \\
&= V\left(T_4 + w_4^F\left(3\right), 4, T_3 + t, 3\right) = 0.727495,
\end{aligned}
$$

$$t_{23}^L (\tau_3) = \min\left(t \,\middle|\, V\left(T_3 + t, 2, T_4 + w_4^F(2), 4\right)\right)$$
$$= V\left(T_3 + t, 3, T_4 + w_4^F(3), 4\right),$$

$$S_{24}^L (\tau_3) = \arg\max_{t \in \left[0, t_{24}^L(\tau_3)\right]} V\left(T_3 + t, 2, T_4 + w_4^F(2), 4\right),$$

$$t_{24}^P = \min\left(t \,\middle|\, V\left(T_2 + t, 2, T_4 + w_4^F(2), 4\right)\right)$$
$$= V\left(T_4 + w_4^F(2), 4, T_2 + t, 2\right) = 1.81706.$$

Table 6.1. Equilibria and type of subgames starting at time T_4 as function of τ_4. Type "P" is preemption game and type "A" is attrition game.

τ_4 region	Type	Leader Technology	Leader Time	Follower Technology	Follower Time
$[0, 0.800591)$	P	4	$T_4 + t_4^P$	4	$T_4 + w_4^F(4)$
$[0.800591, 1.17938)$	A	3	$T_4 + t_{34}^L(\tau_4)$	4	$T_4 + w_4^F(3)$
$[1.17938, 1.87931]$	A	3	$T_4 + S_{34}^L(\tau_4)$	4	$T_4 + w_4^F(3)$
$(1.87931, 1.89322)$	/P	3	$T_4 + t_{34}^P(\tau_4)$	4	$T_4 + w_4^F(3)$
$[1.89322, \infty)$	P	3	T_4	4	$T_4 + w_4^F(3)$

Table 6.2. Equilibria and type of subgames starting at time T_3 as function of τ_3. Type "P" is preemption game and type "A" is attrition game.

τ_3 region	Type	Leader Technology	Leader Time	Follower Technology	Follower Time
$[0, 1.24843)$	P	3	$T_3 + t_{34}^P$	4	$T_4 + w_4^F(3)$
$[1.24843, 2.94586)$	A	2	$T_3 + t_{23}^L(\tau_3)$	4	$T_4 + w_4^F(2)$
$[2.94586, 3.95758)$	A	2	$T_3 + S_{24}^L(\tau_3)$	4	$T_4 + w_4^F(2)$
$[3.95758, \infty)$	A	2	T_3	4	$T_4 + w_4^F(2)$

In Tables 6.2 and 6.3 the equilibrium outcomes are conditional on the next technology not arriving too early. That is the next technology does not arrive before the time at which the leader changes technologies according to the table. In Appendix A the equilibrium outcome functions $\Omega_i(\tau_i)$, $i = 2, 3, 4$, are presented. Let $i \in \{2, 3, 4\}$. In the game that starts in the interval $[T_{i-1}, T_i)$ the equilibrium strategy of a firm depends

Table 6.3. Equilibria and type of subgames starting at time T_2 as function of τ_2. Type "P" is preemption game.

τ_2 region	Type	Leader Technology	Leader Time	Follower Technology	Follower Time
$[0, \infty)$	P	2	$T_2 + t_{24}^P$	4	$T_4 + w_4^F\,(2)$

on τ_i. Note that the higher τ_i the more attractive the technologies $j \in \{1, \dots, i-1\}$ are, due to the decrease of the investment costs of these technologies during the interval $[T_j, T_i)$.

If technology 4 arrives shortly after technology 3 (see first line of Table 6.1), technology 4 dominates technology 3 and both firms will adopt technology 4. If it takes a little longer before technology 4 becomes available, technology 3 is the most attractive technology for the leader to adopt. In the second and third τ_4 region the follower's value is higher than the leader's value. To explain this second mover advantage, consider the second line of Table 6.1. The value of the gain of market share of the follower during the time interval $[T_4 + w_4^F\,(3), \infty)$ outweighs the value of the gain of market share of the leader during the interval $[T_4 + t_{34}^L\,(\tau_4), T_4 + w_4^F\,(3))$. A late arrival of technology 4 makes technology 3 attractive enough for direct adoption, see the last line of Table 6.1. Tables 6.2 and 6.3 should be interpreted in the same way.

We now analyze the game at the moment where technologies 2, 3, and 4 have not been invented yet, in a more elaborate way. Using the outcome function $\Omega_2\,(\tau_2)$ we construct the waiting curve for the game that starts at time T_1 (cf. (6.29)), which is the invention time of the first technology:

$$W\,(t) = \frac{\pi\,(\theta_0, \theta_0)}{r}\,(1 - e^{-rt}) \tag{6.38}$$

$$+ \int_{\tau_2=0}^{\infty} \left[\int_{u=0}^{\tau_2} \pi\,(\theta_0, \theta_0)\,e^{-ru}\,du + e^{-r\tau_2}\Omega_2\,(\tau_2) \right] \lambda e^{-\lambda\tau_2}\,d\tau_2.$$

The leader, follower and joint-moving curves are derived with the equations presented in Section 3. In Figure 6.1 the four curves are plotted.

From Figure 6.1 the following unique ordering of the curves is derived: $F\,(t) > W\,(t) > L\,(t) > M\,(t)$ for all $t \in [T_1, T_2)$. This implies that each firm likes the other to invest as first and does not want to invest as first

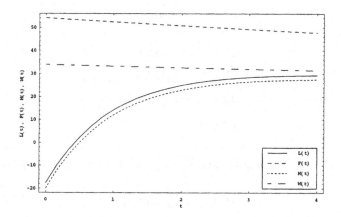

Figure 6.1. Leader, follower, joint moving and waiting curve for the game that starts at time $T_1 = 0$.

itself. Thus waiting is the optimal strategy for the firms in the game that starts in the interval $[T_1, T_2)$.

Then at time T_2 the game starts where technologies 1 and 2 are present, but the remaining technologies 3 and 4 have not been invented yet. From Table 6.3 we derive that one firm will adopt technology 2 at time $T_2 + t_{24}^P$ and the other firm technology 4 if the third technology does not arrive before time $T_2 + t_{24}^P$. With probability

$$\Pr\left(\tau_3 \geq t_{24}^P\right) = e^{-\lambda t_{24}^P} = 0.54570, \tag{6.39}$$

this is the case.

With probability

$$\Pr\left(\tau_3 < t_{24}^P\right) = 1 - e^{-\lambda t_{24}^P} = 0.45430, \tag{6.40}$$

technology 3 arrives before time $T_2 + t_{24}^P$. Now, there are two cases. In the first case, τ_3 is smaller than the boundary 1.24843 (see Table 6.2), which occurs with probability

$$\Pr\left(\tau_3 < 1.24843\right) = 1 - e^{-1.24843\lambda} = 0.34041, \tag{6.41}$$

and in the second case, $1.24843 < \tau_3 < t_{24}^P$, which occurs with probability

$$\Pr\left(1.24843 \leq \tau_3 < t_{24}^P\right) = e^{-1.24843\lambda} - e^{-\lambda t_{24}^P} = 0.11389. \tag{6.42}$$

Table 6.2 states that, in the first case, the outcome will be adoption of technology 3 at time $T_3 + t_{34}^P$ if technology 4 does not arrive before that time. This outcome occurs with the following probability:

$$
\begin{aligned}
& \Pr\left(\tau_3 < t_{24}^P \text{ and } \tau_4 \geq t_{34}^P\right) \\
= \ & \Pr\left(\tau_3 < t_{24}^P\right)\Pr\left(\tau_4 \geq t_{34}^P\right) \\
= \ & \left(1 - e^{-\lambda t_{24}^P}\right)e^{-\lambda t_{34}^P} \\
= \ & 0.26711.
\end{aligned}
\tag{6.43}
$$

Technology 4 arrives before time $T_3 + t_{34}^P$, while $\tau_4 < t_{34}^P$, with probability

$$
\begin{aligned}
& \Pr\left(\tau_3 < t_{24}^P \text{ and } \tau_4 < t_{34}^P\right) \\
= \ & \Pr\left(\tau_3 < t_{24}^P\right)\Pr\left(\tau_4 < t_{34}^P\right) \\
= \ & \left(1 - e^{-\lambda t_{24}^P}\right)\left(1 - e^{-\lambda t_{34}^P}\right) \\
= \ & 0.073303.
\end{aligned}
\tag{6.44}
$$

In this case the outcome will be a preemption equilibrium in which one firm adopts technology 4 at time $T_4 + t_4^P$ and the other firm technology 4 at time $T_4 + w_4^F(4)$. Here it is important to note that $t_{34}^P = 0.727495$ is smaller than the first τ_4 boundary 0.800591. Hence, with probability one the outcomes listed on the lines 2-5 of Table 6.1 will not occur here.

The second case is a little more complicated. The outcome exhibits adoption of technology 3 at time $T_3 + t_{23}^L(\tau_3)$ by one firm, while the other firm adopts technology 4, if technology 4 arrives after time $T_3 + t_{23}^L(\tau_3)$, which happens with probability:

$$
\begin{aligned}
& \Pr\left(1.24843 \leq \tau_3 < t_{24}^P \text{ and } \tau_4 \geq t_{23}^L(\tau_3)\right) \\
= \ & \int_{\tau_3=1.24843}^{t_{24}^P} \Pr\left(\tau_4 \geq t_{23}^L(\tau_3)\right)\lambda e^{-\lambda\tau_3}\,d\tau_3 \\
= \ & \int_{\tau_3=1.24843}^{t_{24}^P} e^{-\lambda t_{23}^L(\tau_3)}\lambda e^{-\lambda\tau_3}\,d\tau_3 \\
= \ & 0.086802.
\end{aligned}
\tag{6.45}
$$

Otherwise the outcome is of the preemption type (first line of Table 6.1) if $\tau_4 < 0.800591$ or of the attrition type (second line of Table 6.1) if $\tau_4 \geq 0.800591$. The probability that the preemption equilibrium occurs is equal to

$$\Pr\left(1.24843 \le \tau_3 < t_{24}^P, \tau_4 < t_{23}^L(\tau_3), \text{ and } \tau_4 < 0.800591\right)$$

$$= \int_{\tau_3=1.24843}^{t_{24}^P} \Pr\left(\tau_4 < \min\left(t_{23}^L(\tau_3), 0.800591\right)\right) \lambda e^{-\lambda \tau_3} d\tau_3$$

$$= 0.026273. \tag{6.46}$$

With probability

$$\Pr\left(1.24843 \le \tau_3 < t_{24}^P, \tau_4 < t_{23}^L(\tau_3), \text{ and } \tau_4 \ge 0.800591\right)$$

$$= \int_{\tau_3=1.24843}^{t_{24}^P} \Pr\left(0.800591 \le \tau_4 < t_{23}^L(\tau_3)\right) \lambda e^{-\lambda \tau_3} d\tau_3$$

$$= \int_{\tau_3=1.24843}^{t_{24}^P} \left(e^{-0.800591\lambda} - e^{-\lambda t_{23}^L(\tau_3)}\right) 1_{\left\{0.800591 \le t_{34}^L(\tau_3)\right\}} \lambda e^{-\lambda \tau_3} d\tau_3$$

$$= 0.00081337, \tag{6.47}$$

the attrition game will happen. Here the leader adopts technology 3 and the follower invests in technology 4. So, on the longer term the follower produces with the more efficient technology which here leads to a higher expected payoff.

The analysis above implies that only the first two lines of Tables 6.1 and 6.2 matter. This for the reason that one of the firms adopts an existing technology, if a new technology arrives too late.

In Table 6.4 all possible outcomes and the probabilities are summarized. We conclude that the ex-ante probability of a preemption equilibrium with rent equalization (see Appendix A) equals 0.91238. The most likely outcome (probability 0.54570) is that one firm adopts technology 2 and the other firm technology 4. With probability 0.087615 there is a second mover advantage in the equilibrium, i.e. the firm that invests as first earns less than the firm that invests as second. The market share gain by the second mover offsets the temporary market share gain of the first mover. With probability 0.90042 the leader adopts another technology than the follower. The follower is expected to adopt technology 4 in all equilibria. Joint adoption does not occur as an equilibrium outcome.

We did not add an extra new technology to the model, because the probability that both firms adopt technology 4 is less than 0.10. Hence, with a probability of more than 0.90 at least one firm invests in another technology than the last one. For this reason we choose not to analyze the game with one technology more.

Table 6.4.   Equilibria and ex-ante probabilities at time $T_1 = 0$. Type "P" is preemption game and type "A" is attrition game.

Probability	Type	Leader Technology	Leader Time	Follower Technology	Follower Time
0.54570	P	2	$T_2 + t_{24}^P$	4	$T_4 + w_4^F$ (2)
0.26711	P	3	$T_3 + t_{34}^P$	4	$T_4 + w_4^F$ (3)
0.086802	A	2	$T_3 + t_{23}^L$ (τ_3)	4	$T_4 + w_4^F$ (2)
0.099576	P	4	$T_4 + t_4^P$	4	$T_4 + w_4^F$ (4)
0.00081337	A	3	$T_4 + t_{34}^L$ (τ_4)	4	$T_4 + w_4^F$ (3)

5.   CONCLUSIONS

We analyzed a framework in which consecutive generations of new technologies arrive over time, and a firm has to make its optimal technology investment decision. Competition on the output market is taken into account. As time passes more efficient technologies arrive according to a stochastic arrival process. The investment cost of a particular technology drops over time.

Introducing the waiting curve as a new concept, the investment decision problem was converted into a timing game. The timing game changes every time a new technology enters the market. We designed an algorithm that can be used to solve this game.

The algorithm is applied to an information technology investment problem with four new technologies. The most likely outcome exhibits diffusion, one firm adopts technology 2 early and the other technology 4 later on, while the expected payoffs of the first and second investor are the same. With a probability of more than 90 percent the expected payoffs of the firms are equal. In the other cases the firm that invests as second performs better than the firm that invests as first. Thus the temporary gain of market share by the leader does not make up for the market share gain of the follower.

One possible extension of this model is to relax the assumption that firms are allowed to make only one technology switch. We believe that this model can be solved in the same fashion: use the waiting curve concept to convert the game to a timing game with multiple actions and solve that game following the work by Simon, 1987b.

Another interesting extension is to make the number of active firms on the output market endogenous. If the active firms make positive profits it may be interesting for a new firm to enter the market. How does the threat of entering change the technology adoption behavior of the existing firms? Will they try to prevent firms to enter the market by adopting new technologies sooner?

Appendix
A. CONSTRUCTION OF THE WAITING CURVE

In this part the waiting curve for the application in Section 3 is constructed. To do so, starting out from each realization the subgames have to be solved. Appendix A of Chapter 4 provides some relevant mathematical prerequisites for the analysis in this appendix.

A.1 GAMES STARTING IN FIFTH PERIOD

The outcome of the subgame that starts at a time after the arrival of the fourth technology depends on the realization of T_4 and thus on the realization of τ_4. It turns out that there are five different intervals for τ_4 to consider. This implies that there are four critical values for τ_4, denoted by $\tau_4^*(i)$, $i \in \{1, 2, 3, 4\}$. On each of these intervals the configuration of the figure in which L, F, and M are depicted is the same.

1. $\tau_4 \in [0, \tau_4^*(1)) = [0, 0.800591)$.

In Figure 6.2 the three graphs of $L(t)$, $F(t)$, and $M(t)$ are plotted for $\tau_4 = 0.5$.

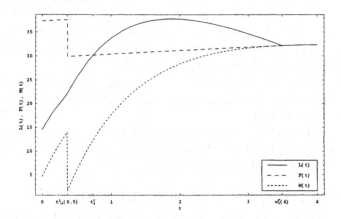

Figure 6.2. Leader, follower and joint moving curves for the subgame that starts at time T_4 if $\tau_4 = 0.5$.

Here technology 4 is invented just after the invention of technology 3. Therefore, technology 4 dominates technology 3 very quickly. At time $t_{34}^L(\tau_4)$ the leader is indifferent between adopting technology 3 and technology 4:

$$V\left(T_4 + t_{34}^L(\tau_4), 3, T_4 + w_4^F(3), 4\right)$$
$$= V\left(T_4 + t_{34}^L(\tau_4), 4, T_4 + w_4^F(4), 4\right),$$

where $w_4^F(3) = 3.26767$ and $w_4^F(4) = 3.49080$. Note that the follower is not indifferent, because the follower curve jumps down at time $t_{34}^L(\tau_4)$. There are two equilibria with symmetric strategies. Define the preemption time t_4^P as

$$t_4^P = \min\left(t \mid L(t) = F(t)\right) = 0.734579.$$

For $t \in \left(t_4^P, w_4^F(4)\right)$ it holds that $L(t) > F(t) > M(t)$. Therefore, the game that starts at time t_4^P is a preemption game.

At $t > t_4^P$ it is in the interest of each firm to adopt technology 4 right away (since $L(t) > F(t)$). But if a firm knows that the other will adopt at this particular time, it wants to preempt at $t - \varepsilon$. Reasoning backwards, at any time beyond t_4^P, firms want to preempt to avoid being preempted later on. As shown in Appendix A of Chapter 4 this leads to the following equilibrium: with probability one-half a firm becomes leader and adopts technology 4 at time $T_4 + t_4^P$. The other firm is follower and adopts technology 4 at time $T_4 + w_4^F(4)$. With probability one-half the roles are reversed. We conclude that the game ends for sure at time $T_4 + t_4^P$. The probability of a mistake, i.e. both firms adopting technology 4 at time $T_4 + t_4^P$ leaving them with a low payoff $M\left(t_4^P\right) < F\left(t_4^P\right)$, is equal to zero (see Appendix A of Chapter 4). Both firms' values are equal, i.e. there is rent-equalization. The firm's value (discounted to time T_4) equals $\frac{1}{2}F\left(t_4^P\right) + \frac{1}{2}L\left(t_4^P\right) = 30.1722$.

At $t < t_4^P$, it holds that $F(t) > L(t)$ and the leader curve is increasing. Therefore, both firms wait until t_4^P where the above described preemption game starts.

The boundary $\tau_4^*(1)$ is derived by solving the equation $t_{34}^L(\tau_4^*(1)) = t_4^P$. Thus if $\tau_4 = \tau_4^*(1)$ the leader is indifferent between the two strategies exactly at the preemption time t_4^P.

2. $\tau_4 \in \left[\tau_4^*(1), \tau_4^*(2)\right) = \left[0.800591, 1.17938\right)$.

The leader, follower and joint-moving curves are plotted in Figure 6.3 for $\tau_4 = 1$.

In this τ_4-region there is no equilibrium with symmetric strategies. Here, there are four equilibria for the subgame. At time $t_{34}^L(\tau_4)$ the

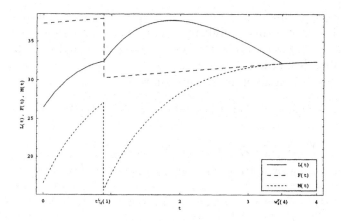

Figure 6.3. Leader, follower and joint moving curves for the subgame that starts at time T_4 if $\tau_4 = 1$.

leader is indifferent between adopting technology 3 and adopting technology 4:

$$V\left(T_4 + t_{34}^L(\tau_4), 3, T_4 + w_4^F(3), 4\right)$$
$$= V\left(T_4 + t_{34}^L(\tau_4), 4, T_4 + w_4^F(4), 4\right).$$

The subgame that starts at time $t > t_{34}^L(\tau_4)$ is a preemption game and in equilibrium the expected value for both firms is

$$V\left(T_4 + w_4^F(4), 4, T_4 + t, 4\right),$$

i.e. the follower value if the leader adopts technology 4 at time $T_4 + t$ and the follower adopts technology 4 at time $T_4 + w_4^F(4)$. This subgame ends at time $T_4 + t$ with probability one.

Adopting before time $T_4 + t_{34}^L(\tau_4)$ is not optimal for a firm, because the follower value is larger than the leader value and the leader value is increasing.

The story above implies that the game will end at time $T_4 + t_{34}^L(\tau_4)$ with probability one. The leader has two possible strategies: adopt technology 3 and adopt technology 4. The follower's optimal reply is always to adopt technology 4. Thus there are two types of equilibria. In the first type the leader adopts technology 3 and the follower technology 4 and in the second type the leader and the follower both adopt technology 4. Right at $T_4 + t_{34}^L(\tau_4)$ the leader's value is equal in both equilibria, but the follower's value is larger in the equilibrium where the leader adopts technology 3. In other words, the equilibrium in which the leader adopts technology 3 Pareto dominates the other equilibrium and that is why we use this equilibrium in further calculations. We assume that nature assigns to a firm the role of leader and that both firms have equal probability of being assigned leader.

The expected value of each firm equals

$$\frac{V\left(T_4+t^L_{34}(\tau_4),3,T_4+w^F_4(3),4\right)+V\left(T_4+w^F_4(3),4,T_4+t^L_{34}(\tau_4),3\right)}{2}.$$

3. $\tau_4 \in \left[\tau^*_4\left(2\right),\tau^*_4\left(3\right)\right] = \left[1.17938, 1.87931\right].$

The leader curve is decreasing on the interval $\left(S^L_{34}\left(\tau_4\right),t^L_{34}\left(\tau_4\right)\right)$, with

$$S^L_{34}\left(\tau_4\right) = \arg\max_{t\in\left[0,t^L_{34}(\tau_4)\right]} V\left(T_4+t,3,T_4+w^F_4\left(3\right),4\right).$$

The boundary $\tau^*_4\left(2\right)$ is the solution of the equation

$$S^L_{34}\left(\tau^*_4\left(2\right)\right) = t^L_{34}\left(\tau^*_4\left(2\right)\right).$$

In Figure 6.4 the three curves are plotted for $\tau_4 = 1.5$.

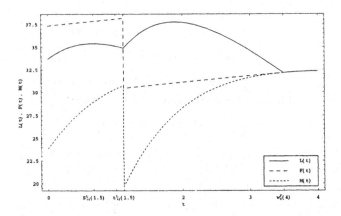

Figure 6.4.   Leader, follower and joint moving curves for the subgame that starts at time T_4 if $\tau_4 = 1.5$.

Since the follower curve lies above the leader curve, we have an attrition game on this interval, because for all $t \in \left[0,t^L_{34}\left(\tau_4\right)\right]$:

$$V\left(T_4+t,3,T_4+w^F_4\left(3\right),4\right) > V\left(T_4+w^F_4\left(4\right),4,T_4+t^L_{34}\left(\tau_4\right),4\right),$$

there does not exist an equilibrium with symmetric strategies for the game (cf. Appendix A.3 of Chapter 4). There are two equilibria for this game. In each equilibrium, the leader adopts technology 3 at time $T_4 + S_{34}^L (\tau_4)$ and the follower adopts technology 4 at time $T_4 + w_4^F (3)$. As before we assume that nature assigns a firm to be leader or follower. Both firms have equal probability of being assigned leader. Thus the expected value of a firm equals

$$\frac{V\left(T_4 + S_{34}^L(\tau_4), 3, T_4 + w_4^F(3), 4\right) + V\left(T_4 + w_4^F(3), 4, T_4 + S_{34}^L(\tau_4), 3\right)}{2}.$$

The subgames that start at time $t > t_{34}^L (\tau_4)$ have not changed.

4. $\tau_4 \in \left(\tau_4^* (3), \tau_4^* (4)\right) = (1.87931, 1.89322)$.

In these subgames the value of the leader exceeds the value of the follower during a part of the interval $\left(0, t_{34}^L (\tau_4)\right)$, see Figure 6.5.

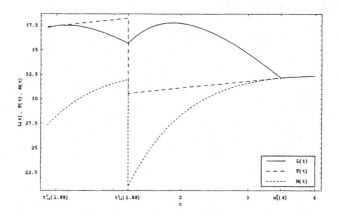

Figure 6.5. Leader, follower and joint moving curves for the subgame that starts at time T_4 if $\tau_4 = 1.88$.

These subgames are preemption games. There are two equilibria with symmetric strategies. With probability one-half a firm becomes leader and adopts technology 3 at time $T_4 + t_{34}^P (\tau_4)$, where $t_{34}^P (\tau_4)$ is defined by

$$\begin{aligned} t_{34}^P (\tau_4) &= \min \left(t \, | V \left(T_4 + t, 3, T_4 + w_4^F (3), 4\right)\right) \\ &= V \left(T_4 + w_4^F (3), 4, T_4 + t, 3\right). \end{aligned}$$

The other firm is follower and adopts technology 4 at time $T_4 + w_4^F (3)$, and with probability one-half the roles are reversed. According to Appendix A of Chapter 4, due to rent equalization, there is zero probability of mistake, i.e. both firms adopting technology 3 at time $T_4 + t_{34}^P (\tau_4)$.

Both firm's values are equal so that there is rent-equalization. The firm's value (discounted to time T_4) equals

$$V \left(T_4 + t_{34}^P (\tau_4), 3, T_4 + w_4^F (3), 4 \right).$$

The boundary $\tau_4^* (3)$ is defined as the smallest τ_4 for which there exists an $t_{34}^P (\tau_4)$.

5. $\tau_4 \in \left[\tau_4^* (4), \infty \right) = [1.89322, \infty)$.

The boundary $\tau_4^* (4)$ is defined as the smallest τ_4 for which the pre-emption time $t_{34}^P (\tau_4)$ equals 0. Thus, in this region the games end at time T_4 with probability one. The leader's value at time T_4 exceeds the follower's value at time T_4 and that is why there is a positive probability of a mistake, see Figure 6.6.

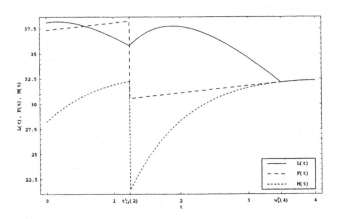

Figure 6.6. Leader, follower and joint moving curves for the subgame that starts at time T_4 if $\tau_4 = 2$.

Define (see Appendix A of Chapter 4)

$$\alpha (t \, | \tau_4) = \frac{V \left(T_4 + t, 3, T_4 + w_4^F (3), 4 \right) - V \left(T_4 + w_4^F (3), 4, T_4 + t, 3 \right)}{V \left(T_4 + t, 3, T_4 + w_4^F (3), 4 \right) - V \left(T_4 + t, 3, T_4 + t, 3 \right)}.$$

The probability of a firm to become leader (adopt technology 3 at time T_4) or to become follower (adopt technology 4 at time $T_4 + w_4^F (3)$) equals

$$\frac{1 - \alpha (0 \, | \tau_4)}{2 - \alpha (0 \, | \tau_4)},$$

and the probability of a mistake (both firms adopting technology 3 at time T_4) equals

$$\frac{\alpha\left(0\,|\tau_4\right)}{2-\alpha\left(0\,|\tau_4\right)}.$$

Using these probabilities, it is not hard to derive that the expected value of the firm equals

$$V\left(T_4 + w_4^F\left(3\right), 4, T_4, 3\right).$$

SUMMARY

The expected payoff $\Omega_4\left(\tau_4\right)$ of the game equals

$$V\left(T_4 + w_4^F\left(4\right), 4, T_4 + t_4^P, 4\right), \qquad (6.48)$$

if $\tau_4 \in [0, \tau_4^*\left(1\right))$,

$$\frac{V\left(T_4 + t_{34}^L(\tau_4), 3, T_4 + w_4^F(3), 4\right) + V\left(T_4 + w_4^F(3), 4, T_4 + t_{34}^L(\tau_4), 3\right)}{2}, \qquad (6.49)$$

if $\tau_4 \in [\tau_4^*\left(1\right), \tau_4^*\left(2\right))$,

$$\frac{V\left(T_4 + S_{34}^L(\tau_4), 3, T_4 + w_4^F(3), 4\right) + V\left(T_4 + w_4^F(3), 4, T_4 + S_{34}^L(\tau_4), 3\right)}{2}, \qquad (6.50)$$

if $\tau_4 \in [\tau_4^*\left(2\right), \tau_4^*\left(3\right)]$,

$$V\left(T_4 + t_{34}^P\left(\tau_4\right), 3, T_4 + w_4^F\left(3\right), 4\right), \qquad (6.51)$$

if $\tau_4 \in (\tau_4^*\left(3\right), \tau_4^*\left(4\right))$, and

$$V\left(T_4 + w_4^F\left(3\right), 4, T_4, 3\right), \qquad (6.52)$$

if $\tau_4 \in [\tau_4^*\left(4\right), \infty)$.

A.2 GAME STARTING IN FOURTH PERIOD

Using the expressions (6.48)-(6.52) for $\Omega_4\left(\tau_4\right)$ we derive the waiting curve for the subgames starting at some time $t \in [T_3, T_4)$:

$$W\left(t\right) = \frac{\pi\left(\theta_0, \theta_0\right)}{r}\left(1 - e^{-rt}\right) + e^{-rt}$$

$$\times \int_{u_3=0}^{\infty}\left[\int_{v_3=0}^{u_3}\pi\left(\theta_0, \theta_0\right)e^{-rv_3}\,dv_3 + e^{-ru_3}\Omega_4\left(u_3\right)\right]\lambda e^{-\lambda u_3}\,du_3.$$

The equilibria in this subgame depend on τ_3. There are four different τ_3 intervals to consider. Thus there are three critical values for τ_3: $\tau_3^*\left(i\right)$, $i \in \{1, 2, 3\}$. For the moment we derive the equilibria in the case that technology 4 has not arrived yet.

1. $\tau_3 \in [0, \tau_3^* (1)) = [0, 1.24843)$.

In Figure 6.7 the leader, follower, joint moving and waiting curves are plotted for $\tau_3 = 1$. $t_{34}^F (= 3.24608)$ is defined as the point in time at which the follower is indifferent between adopting technology 4 at time $T_4 + w_4^F (3)$ and adopting technology 3 at time $T_3 + w_3^F (3)$ given that the leader adopted technology 3 and that technology 4 has not arrived yet.

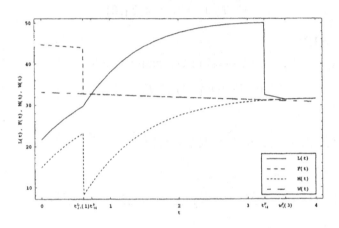

Figure 6.7. Leader, follower, joint moving and waiting curve for the subgame that starts at time T_3 if $\tau_3 = 1$.

It turns out that waiting is not an option, because the leader curve exceeds the waiting curve for some points in time and the corresponding timing game without waiting curve is a preemption game. As usual there are two equilibria with symmetric strategies. With probability one-half a firm becomes leader and adopts technology 3 at time $T_3 + t_{34}^P$, where $t_{34}^P = 0.727495$. The other firm is follower and is expected to adopt technology 4 at time $T_4 + w_4^F (3)$, and with probability one-half the roles are reversed. There is zero probability of mistake, i.e. both firms adopting technology 3 at time $T_3 + t_{34}^P$. Both firm's values are equal so that there is rent-equalization. The firm's value (discounted to time T_3) equals 32.6639.

If $\tau_3 = \tau_3^* (1) = 1.24843$ it holds that $t_{23}^L (\tau_3) = t_{34}^P$.

2. $\tau_3 \in [\tau_3^* (1), \tau_3^* (2)) = [1.24843, 2.94586)$.

In this region there are two types of equilibria, but none of them is supported by symmetric strategies. In Figure 6.8 the four curves are plotted for $\tau_3 = 2$.

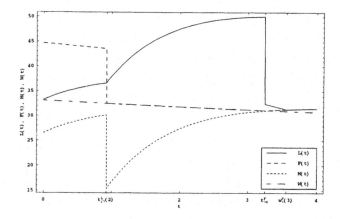

Figure 6.8. Leader, follower, joint moving and waiting curve for the subgame that starts at time T_3 if $\tau_3 = 2$.

We use the following type in further calculations. The leader adopts technology 2 at time $T_3 + t_{23}^L (\tau_3)$ and the follower is expected to wait for technology 4 and adopt it at time $T_4 + w_4^F (2)$, where $w_4^F (2) = 2.97998$. Nature assigns the roles to the firms. The expected value of the firms is equal to

$$\frac{V\left(T_3 + t_{23}^L(\tau_3), 2, T_4 + w_4^F(2), 4\right) + V\left(T_4 + w_4^F(2), 4, T_3 + t_{23}^L(\tau_3), 2\right)}{2}.$$

The second boundary, $\tau_3^* (2) = 2.94586$, is derived by solving the following equation

$$t_{23}^L (\tau_3) = S_{24}^L (\tau_3),$$

where

$$S_{24}^L (\tau_3) = \arg \max_{t \in \left[0, t_{23}^L(\tau_3)\right]} V \left(T_3 + t, 2, T_4 + w_4^F (2), 4\right).$$

3. $\tau_3 \in \left[\tau_3^* (2), \tau_3^* (3)\right) = [2.94586, 3.95758)$.

Again no equilibrium with symmetric strategies in this region exists, see Figure 6.9 for a plot of the curves in this region.

In equilibrium the leader adopts technology 2 at time $T_3 + S_{24}^L (\tau_3)$ and the follower is expected to adopt technology 4 at time $T_4 + w_4^F (2)$. As before the roles are assigned by nature. The firm's expected value equals

$$\frac{V\left(T_3 + S_{24}^L(\tau_3), 2, T_4 + w_4^F(2), 4\right) + V\left(T_4 + w_4^F(2), 4, T_3 + S_{24}^L(\tau_3), 2\right)}{2}.$$

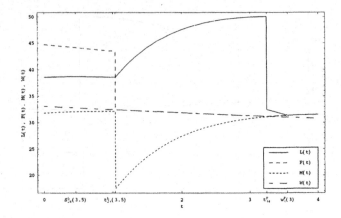

Figure 6.9. Leader, follower, joint moving and waiting curve for the subgame that starts at time T_3 if $\tau_3 = 3.5$.

The critical value $\tau_3^* (3)$ ($= 3.95758$) is defined by

$$\tau_3^* (3) = \min \left(\tau_3 |\, S_{24}^L (\tau_3) = 0 \right).$$

4. $\tau_3 \in \left[\tau_3^* (3) , \infty \right) = [3.95758, \infty) .$

In Figure 6.10 the leader, follower, joint moving and waiting curves are plotted for $\tau_3 = 4$.

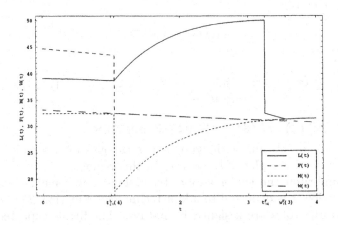

Figure 6.10. Leader, follower, joint moving and waiting curve for the subgame that starts at time T_3 if $\tau_3 = 4$.

The leader adopts technology 2 at time T_3 and the follower is expected to adopt technology 4 at time $T_4 + w_4^F(2)$ in equilibrium. Roles are assigned by nature. Expected firm values are given by

$$\frac{1}{2}\left(V\left(T_3, 2, T_4 + w_4^F(2), 4\right) + V\left(T_4 + w_4^F(2), 4, T_3, 2\right)\right).$$

SUMMARY

The expected payoff of the game $\Omega_3(\tau_3)$ equals

$$\int_{\tau_4=0}^{t_{34}^P}\left[\int_{v=0}^{\tau_4}\pi(\theta_0,\theta_0)e^{-rv}dv + e^{-r\tau_4}\Omega_4(\tau_4)\lambda e^{-\lambda\tau_4}\right]d\tau_4$$

$$+\int_{\tau_4=t_{34}^P}^{\infty} V\left(T_4 + w_4^F(3), 4, T_3 + t_{34}^P, 3\right)\lambda e^{-\lambda\tau_4}d\tau_4, \qquad (6.53)$$

if $\tau_3 \in [0, \tau_3^*(1))$. If $\tau_3 \in [\tau_3^*(1), \tau_3^*(2))$, $\Omega_3(\tau_3)$ equals

$$\int_{\tau_4=0}^{t_{23}^L(\tau_3)}\int_{v=0}^{\tau_4}\left[\pi(\theta_0,\theta_0)e^{-rv}dv + e^{-r\tau_4}\Omega_4(\tau_4)\lambda e^{-\lambda\tau_4}\right]d\tau_4$$

$$+\frac{1}{2}\int_{\tau_4=t_{23}^L(\tau_3)}^{\infty} V\left(T_3 + t_{23}^L(\tau_3), 2, T_4 + w_4^F(2), 4\right)\lambda e^{-\lambda\tau_4}d\tau_4$$

$$+\frac{1}{2}\int_{\tau_4=t_{23}^L(\tau_3)}^{\infty} V\left(T_4 + w_4^F(2), 4, T_3 + t_{23}^L(\tau_3), 2\right)\lambda e^{-\lambda\tau_4}d\tau_4. \qquad (6.54)$$

If $\tau_3 \in [\tau_3^*(2), \tau_3^*(3))$, $\Omega_3(\tau_3)$ is given by

$$\int_{\tau_4=0}^{S_{24}^L(\tau_3)}\left[\int_{v=0}^{\tau_4}\pi(\theta_0,\theta_0)e^{-rv}dv + e^{-r\tau_4}\Omega_4(\tau_4)\lambda e^{-\lambda\tau_4}\right]d\tau_4$$

$$+\frac{1}{2}\int_{\tau_4=S_{24}^L(\tau_3)}^{\infty} V\left(T_3 + S_{24}^L(\tau_3), 2, T_4 + w_4^F(2), 4\right)\lambda e^{-\lambda\tau_4}d\tau_4$$

$$+\frac{1}{2}\int_{\tau_4=S_{24}^L(\tau_3)}^{\infty} V\left(T_4 + w_4^F(2), 4, T_3 + S_{24}^L(\tau_3), 2\right)\lambda e^{-\lambda\tau_4}d\tau_4. \qquad (6.55)$$

and

$$\Omega_3\left(\tau_3\right) = \frac{1}{2}\left(V\left(T_3, 2, T_4 + w_4^F\left(2\right), 4\right) + V\left(T_4 + w_4^F\left(2\right), 4, T_3, 2\right)\right),$$
(6.56)

if $\tau_3 \in \left[\tau_3^*\left(3\right), \infty\right).$

A.3   GAMES STARTING IN THIRD PERIOD

Using the expressions (6.53)-(6.56) for $\Omega_3\left(\tau_3\right)$ we derive the waiting curve for the subgames starting at some time $t \in \left[T_2, T_3\right)$:

$$
\begin{aligned}
W\left(t\right) = & \frac{\pi\left(\theta_0, \theta_0\right)}{r}\left(1 - e^{-rt}\right) + e^{-rt} \\
& \times \int_{u_2=0}^{\infty}\left[\int_{v_2=0}^{u_2}\pi\left(\theta_0, \theta_0\right)e^{-rv_2}dv_2 + e^{-ru_2}\Omega_3\left(u_2\right)\right]\lambda e^{-\lambda u_2}du_2.
\end{aligned}
$$

It turns out that the equilibria in this subgames do not depend on τ_2. This can be explained by the fact that $t_{12}^L\left(\tau_2\right) < t_{24}^P$ for all $\tau_2 \in \left[0, \infty\right).$ In Figure 6.11 the four curves are plotted for $\tau_2 = 2$. $t_{12}^L\left(\tau_2\right)$ is the point in time from T_2 on at which the leader is indifferent between adopting technology 1 and adopting technology 2 given that technology 3 has not arrived yet and the follower adopts technology 4 at time $T_4 + w_4^F\left(1\right)$ and $T_4 + w_4^F\left(2\right)$, respectively.

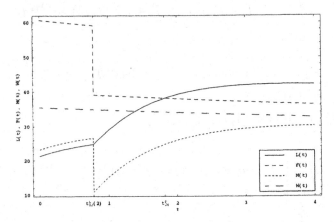

Figure 6.11. Leader, follower, joint moving and waiting curve for the subgame that starts at time T_2 if $\tau_2 = 2$.

There are two equilibria, where each occurs with a probability one-half. One firm is the leader and adopts technology 2 at time $T_2 + t_{24}^P$, where $t_{24}^P = 1.81706$, and the other one is the follower and is expected to adopt technology 4 at time $T_4 + w_4^F(2)$. The firm's expected value equals 38.0414.

SUMMARY

The expected payoff of the game $\Omega_2(\tau_2)$ equals for $\tau_2 \in [0, \infty)$,

$$\int_{u_2=0}^{t_{24}^P} \left[\int_{v_2=0}^{u_2} \pi(\theta_0, \theta_0) e^{-rv_2} dv_2 + e^{-ru_2} \Omega_3(u_2) \lambda e^{-\lambda u_2} \right] du_2$$

$$+ \int_{u_2=t_{24}^P}^{\infty} V\left(T_4 + w_4^F(2), 4, T_2 + t_{24}^P, 2\right) \lambda e^{-\lambda u_2} du_2.$$

where $z_n^* = 1.81705$, and the other one is the follower and is expected to adopt technology A at time $t_A^* = t_A^*(2)$. The firm's expected value equals 58.0114.

SUMMARY

The expected payoff of the firm ... equals for $r_2 \in [0, \infty)$:

III

GAME THEORETIC REAL OPTION MODELS

Chapter 7

ONE NEW TECHNOLOGY
AND SYMMETRIC FIRMS

1.   INTRODUCTION

This chapter considers a framework with two identical firms which both have the possibility to make an investment that increases their payoff. By how much this payoff is raised is not known beforehand, since the future market conditions for the firm's products are uncertain. Both firms operate on the same output market which implies that the investment decision of one firm affects the payoff of the other firm. By analyzing this model uncertainty is combined with strategic aspects.

We identify three scenarios. In the first scenario a preemption equilibrium occurs, where the moments of investment of both firms are dispersed. The first scenario particularly holds when first mover advantages are large. In the second scenario the outcome is that the firms simultaneously invest at the moment that demand is relatively large. In the third scenario it turns out that in economic environments with low uncertainty the preemption equilibrium is applied, while with large uncertainty both firms invest together at the moment that demand is large. This is understandable since the option value of waiting rises with uncertainty. Then opportunity costs of investment are large so that the output market conditions must compensate for this when the firm invests.

Furthermore we find that, compared to the monopoly situation, the demand trigger value is lower for the first investor in the preemption equilibrium. Hence, in order to be able to preempt its rival, the firm is satisfied with a lower revenue at the moment it invests. Therefore, the discounted cash flow stream of the investment, which equals the strategic option value of waiting, is lower than the option value of waiting that prevails in a monopoly situation. On the other hand, the demand trigger value in the joint investment case is higher than in the monopoly case. The reason is that the market has to be shared by two firms. It turns out that in the joint investment case the strategic option value of waiting exactly equals the option value of waiting for the monopoly case.

Finally we compare our analysis with the few contributions that include the real option framework in multiple firm models. Doing this we are able to make a methodological point. In the preemption equilibrium, situations occur where it is optimal for one firm to invest, but at the same time investment is not beneficial if both firms decide to do so. Nevertheless, since the firms are identical there is a possibility that still both firms invest at the same time, which leads to a low payoff for both of them. Following the approach described in Appendix A of Chapter 4 we obtain that such a coordination failure can occur with positive probability at moments of time where the leader's payoff is strictly larger than the follower's payoff. Most contributions in this area, such as Grenadier, 1996, Dutta et al., 1995, and Weeds, 1999, make unsatisfactory assumptions with the aim to be able to ignore the possibility of simultaneous investment at points of time that this is not optimal. Grenadier assumes that "if each (firm) tries to build first, one will randomly (i.e., through the toss of a coin) win the race", see pp. 1656-1657 of Grenadier, 1996, while in Dutta et al., 1995 on p. 568, it is assumed that "if both (firms) i and j attempt to enter at any period t, then only one of them succeeds in doing so".

The model is presented in Section 2. In Section 3 we solve the investment problem if there is only one firm active. This will give the benchmark result. The duopoly model is solved in Section 4. In Section 5 comparisons are made with related contributions. Section 6 concludes.

2.   THE MODEL

Two identical firms are active on a market and have the possibility to make an irreversible investment which results in a higher profit flow. A possible interpretation is that both firms have the possibility to adopt a new technology which after adoption increases the firm's profit. We assume that the firms are risk neutral, value maximizing and discount with constant factor $r\,(>0)$. The sunk cost to adopt the new technology is constant and equals $I\,(>0)$. Future profits are of a yet unknown size. When we denote one firm by i, the other firm is denoted by j, with $i, j \in \{1, 2\}$ and $i \neq j$.

At time $t\,(\geq 0)$ the profit flow of firm i equals

$$Y(t)\, D_{N_i N_j},\qquad\qquad (7.1)$$

where, for $k \in \{i, j\}$:

$$N_k = \begin{cases} 0 & \text{if firm } k \text{ has not invested,} \\ 1 & \text{if firm } k \text{ has invested.} \end{cases} \tag{7.2}$$

In order to incorporate uncertainty, $Y(t)$ follows a geometric Brownian motion process:

$$dY(t) = \mu Y(t)\, dt + \sigma Y(t)\, d\omega(t), \tag{7.3}$$
$$Y(0) = y, \tag{7.4}$$

where $y > 0$, $0 < \mu < r$, $\sigma > 0$, and the $d\omega(t)$'s are independently and identically distributed according to a normal distribution with mean zero and variance dt. Keeping in mind that (i) the irreversible investment increases the profit flow and (ii) the firm obtains higher profits if the competitor is weak (thus not having invested (yet)), the following restrictions on $D_{N_i N_j}$ are implied:

$$D_{10} > D_{11} > D_{00} > D_{01}. \tag{7.5}$$

Further we assume that there is a first mover advantage to investment:

$$D_{10} - D_{00} > D_{11} - D_{01}. \tag{7.6}$$

Note that, contrary to Nielsen, 1999, we only consider the case in which there are negative externalities to investment. That is, it is better for the firm that the other firm has not invested ($D_{10} > D_{11}$). When $D_{11} > D_{10}$ there are positive externalities to investment, which can be caused by, e.g., network externalities or the fact that firms produce complementary products. The aim of this chapter is to study effects of strategic interactions on the option value of waiting, and thus on the speed of investment.

3. MONOPOLY

In this section we assume that there is only one firm active on the output market. We use the solution of this model as a benchmark for the results of the duopoly model.

From here on we omit the time dependence of Y, whenever confusion is not possible. The problem facing the firm is an optimal stopping problem (see also Appendix A of Chapter 2). Hence, intuition suggests that there exists a threshold Y_M such that investing is optimal if $Y \geq Y_M$ and waiting is optimal when $Y < Y_M$. Denote the value of the firm at Y before the investment by $V(Y)$. In Appendix A.1 we derive that

$$V(Y) = \begin{cases} A_1 Y^{\beta_1} + \frac{Y D_{00}}{r - \mu} & \text{if } Y < Y_M, \\ \frac{Y D_{10}}{r - \mu} - I & \text{if } Y \geq Y_M, \end{cases} \tag{7.7}$$

where β_1 is the positive root of the following quadratic equation

$$\frac{1}{2}\sigma^2\beta^2 + \left(\mu - \frac{1}{2}\sigma^2\right)\beta - r = 0. \tag{7.8}$$

Expressions for the investment threshold Y_M and the constant A_1 are found by exploiting the so called value matching and smooth pasting conditions (see Appendix A of Chapter 2):

$$A_1 Y_M^{\beta_1} + \frac{Y_M D_{00}}{r - \mu} = \frac{Y_M D_{10}}{r - \mu} - I, \tag{7.9}$$

$$\beta_1 A_1 Y_M^{\beta_1 - 1} + \frac{D_{00}}{r - \mu} = \frac{D_{10}}{r - \mu}. \tag{7.10}$$

Solving the last two equations gives

$$Y_M = \frac{\beta_1}{\beta_1 - 1} \frac{(r - \mu) I}{D_{10} - D_{00}}, \tag{7.11}$$

$$A_1 = \frac{Y_M^{1 - \beta_1}}{\beta_1} \frac{D_{10} - D_{00}}{r - \mu}. \tag{7.12}$$

The optimal investment strategy of the firm is to invest at time T_M, where

$$T_M = \inf\left(t \mid Y(t) \geq Y_M\right). \tag{7.13}$$

The following proposition states that the threshold Y_M is unique. The proof is given in Appendix B.

PROPOSITION 7.1 *The threshold Y_M (see equation (7.11)) is unique.*

When Y is below the threshold value Y_M the value of the firm consists of two parts (see expression (7.7)). The first part resembles the value of the option to invest and the second part is the expected value of the firm if the firm never invests. The option value rises with uncertainty (β_1 is decreasing in σ and note (7.7) and (7.12)), thus uncertainty creates value for the firm. The implication is that the investment threshold also rises with uncertainty, so that the firm's willingness to invest decreases with uncertainty. Intuitively this can be understood by noting that under large uncertainty it is more valuable to wait for new information about the profitability of an investment before undertaking it. As stressed in Dixit and Pindyck, 1996 the difference between the traditional net present value method and the real options approach to investment problems is completely captured in the factor $\frac{\beta_1}{\beta_1 - 1}$ (> 1), that occurs in the threshold value (see (7.11)). The net present value would be equal to zero if the firm would invest when $Y = \frac{(r-\mu)I}{D_{10}-D_{00}}$. Investing when $Y = Y_M$ thus gives a positive net present value:

$$\frac{Y_M D_{10}}{r - \mu} - I - \frac{Y_M D_{00}}{r - \mu} = \frac{I}{\beta_1 - 1} > 0. \tag{7.14}$$

From the theory of financial options we know that it is only optimal to exercise an option if it is sufficiently deep *in the money*, whereas the net present value method prescribes to exercise the investment option when it is *at the money*.

4.   DUOPOLY

In this section we extend the model of Section 3 by adding another identical firm. We solve the model in which both firms are initially active on the output market. This distinguishes our model from Chapter 9 of Dixit and Pindyck, 1996, where the firms do not produce initially. Then a firm enters a new market at the moment that it invests. Note that this new market model is retrieved by setting

$$D_{00} = D_{01} = 0. \tag{7.15}$$

We compare our results to those of the new market model in Section 5.

We call the firm that invests first the leader, and the other firm is the follower. The model is solved backwards. First we derive the optimal investment decision for the follower, and using that we derive the optimal investment strategy for the leader. In Subsection 4.3 the optimal joint investment outcome is derived. The analysis of the first three subsections is used in Subsection 4.4, where we characterize the possible equilibria. In Subsection 4.5 we describe the properties of the equilibria and compare them with the outcome of the monopoly model.

4.1   FOLLOWER

For the moment let us assume that the leader has invested. In the same way as $V(Y)$ is derived in (7.7), the value of the follower is derived and is given by

$$F(Y) = \begin{cases} B_1 Y^{\beta_1} + \frac{Y D_{01}}{r - \mu} & \text{if } Y < Y_F, \\ \frac{Y D_{11}}{r - \mu} - I & \text{if } Y \geq Y_F. \end{cases} \tag{7.16}$$

The threshold Y_F is defined in the same fashion as Y_M: it is the point at which the follower is indifferent between investing and not investing. In the same way as the proof of Proposition 7.1 one can prove that the threshold Y_F is unique.

When Y is smaller than Y_F the value of the follower equals the value of the option to invest, $B_1 Y^{\beta_1}$, plus the value of never investing, $\frac{Y D_{01}}{r - \mu}$. Solving the value matching and smooth pasting conditions gives

$$B_1 = \frac{Y_F^{1-\beta_1}}{\beta_1} \frac{D_{11} - D_{01}}{r - \mu}, \tag{7.17}$$

$$Y_F = \frac{\beta_1}{\beta_1 - 1} \frac{(r - \mu) I}{D_{11} - D_{01}}. \tag{7.18}$$

Due to equation (7.5) the last two expressions are strictly positive. It is optimal for the follower to invest at time T_F, where

$$T_F = \inf \left(t \,|\, Y(t) \geq Y_F \right). \tag{7.19}$$

4.2   LEADER

The expected value of the leader at time t when he invests at time $t \, (< T_F)$ equals

$$L(Y(t)) = E \left[\int_{\tau=t}^{T_F} Y(\tau) D_{10} e^{-r(\tau-t)} d\tau - I \right.$$

$$\left. + \int_{\tau=T_F}^{\infty} Y(\tau) D_{11} e^{-r(\tau-t)} d\tau \right]. \tag{7.20}$$

Working out the expectation (see Appendix A.2 for details) gives

$$L(Y) = \frac{Y D_{10}}{r - \mu} - I + \left(\frac{Y}{Y_F} \right)^{\beta_1} \frac{Y_F (D_{11} - D_{10})}{r - \mu}. \tag{7.21}$$

If the leader invests when $Y \geq Y_F$, the follower will invest too, so that the leader's expected value equals the value of joint investment, denoted by $M(Y)$:

$$M(Y) = \frac{Y D_{11}}{r - \mu} - I. \tag{7.22}$$

4.3   JOINT INVESTMENT

We assume that the firms invest simultaneously at time T_θ, where

$$T_\theta = \inf \left(t \,|\, Y(t) \geq \theta \right), \tag{7.23}$$

for some $\theta > 0$. The expected value of each firm at time $t\,(< T_\theta)$ equals

$$
J\left(Y\left(t\right),\theta\right) \;=\; E\left[\int_{\tau=t}^{T_\theta} Y\left(\tau\right) D_{00} e^{-r(\tau-t)} d\tau - I e^{-r(T_\theta-t)} \right.
$$

$$
\left. + \int_{\tau=T_\theta}^{\infty} Y\left(\tau\right) D_{11} e^{-r(\tau-t)} d\tau \right]. \tag{7.24}
$$

Thus

$$
J\left(Y,\theta\right) = \begin{cases} \frac{Y D_{00}}{r-\mu} + \left(\frac{Y}{\theta}\right)^{\beta_1}\left(\frac{\theta(D_{11}-D_{00})}{r-\mu} - I\right) & \text{if } Y < \theta, \\ \frac{Y D_{11}}{r-\mu} - I & \text{if } Y \geq \theta. \end{cases} \tag{7.25}
$$

Note that $M\left(Y\right) = J\left(Y,Y\right)$. The optimal joint investment time T_J equals

$$
T_J = \inf\left(t \mid Y\left(t\right) \geq Y_J\right), \tag{7.26}
$$

where Y_J is given by (analogous to (7.11)):

$$
Y_J = \frac{\beta_1}{\beta_1 - 1} \frac{(r-\mu) I}{D_{11} - D_{00}}. \tag{7.27}
$$

Analogous to the Y_M it can be proved that Y_J is unique.

4.4   EQUILIBRIA

It turns out to be convenient to distinguish between the following two cases. In the first case there exists a Y such that there are incentives to become the leader. With other words, for such a Y the leader's payoff, $L\left(Y\right)$, which can be obtained after investing right away, exceeds the joint investment payoff which the firm obtains when it waits with investment until Y reaches Y_J, at which both firms invest simultaneously. This implies that

$$
\exists Y \in (0, Y_F) \text{ such that } L\left(Y\right) > J\left(Y,Y_J\right). \tag{7.28}
$$

In the second case there does not exist such a Y, so that

$$
L\left(Y\right) \leq J\left(Y,Y_J\right) \text{ for all } Y \in (0, Y_F). \tag{7.29}
$$

Since the firms are identical, no reason can be found why they should behave differently. Therefore, we concentrate on equilibria that are supported by symmetric strategies. We use the perfect equilibrium concept for timing games that is described in Appendix A of Chapter 4. There it is argued that in this kind of games a strategy can not be represented by a single distribution function. It is necessary to be able to distinguish between types of atoms. Therefore the closed loop strategy of firm i consists of a collection of simple strategies: $\left(G_i^t\left(\cdot\right), \alpha_i^t\left(\cdot\right)\right)_{t \geq 0}$. The time index t denotes the starting time of the game. $G_i^t\left(s\right)$ is the probability that firm i has invested by some time s given that the other firm has not invested. The function $\alpha_i^t\left(s\right)$ measures the intensity of atoms on the interval $[s, s + ds]$. By definition $\alpha_i^t\left(s\right) > 0$ implies that a firm is sure to invest by time s, i.e. $G_i^t\left(s\right) = 1$.

Next we give an interpretation of the function $\alpha_i^t\left(s\right)$. Forget for a moment the dependence on t. Let τ_i be the smallest point in time at which $\alpha_i\left(s\right)$ is positive: $\tau_i = \inf\left\{s \mid \alpha_i\left(s\right) > 0\right\}$ and define τ to be equal to $\tau = \min\left(\tau_1, \tau_2\right)$. From the definition we know for sure that at least one firm has invested by time τ.

The function value $\alpha_1\left(\tau\right)\left(\alpha_2\left(\tau\right)\right)$ should be interpreted as the probability that firm 1 (2) chooses row (column) 1 in the matrix game of which the payoffs are depicted in Figure 7.1. Playing the game costs no time and if player 1 chooses row 2 and player 2 column 2 the game is repeated. If necessary the game will be repeated infinitely often.

	$\alpha_2\left(\tau\right)$ firm 2	$1-\alpha_2\left(\tau\right)$
$\alpha_1\left(\tau\right)$	$(\text{M}(\text{Y}(\tau)), \text{M}(\text{Y}(\tau)))$	$(\text{L}(\text{Y}(\tau)), \text{F}(\text{Y}(\tau)))$
firm 1		
$1-\alpha_1\left(\tau\right)$	$(\text{F}(\text{Y}(\tau)), \text{L}(\text{Y}(\tau)))$	repeat game

Figure 7.1. Payoffs and strategies for firm 1 and firm 2 of the matrix game played at time τ.

In our model the firms will use the same strategy, so that $\alpha^t(s) = \alpha_i^t(s) = \alpha_j^t(s)$. Then the probability that firm i is the only firm that invests at time τ, $\Pr(\text{one}|\tau)$, equals

$$\Pr(\text{one}|\tau) = \alpha^t(\tau)\left(1 - \alpha^t(\tau)\right) + \left(1 - \alpha^t(\tau)\right)\left(1 - \alpha^t(\tau)\right)\Pr(\text{one}|\tau),$$

which gives

$$\Pr(\text{one}|\tau) = \frac{1 - \alpha^t(\tau)}{2 - \alpha^t(\tau)}. \tag{7.30}$$

For the probability that both firms invest at τ, $\Pr(\text{two}|\tau)$, we get

$$\Pr(\text{two}|\tau) = \alpha^t(\tau)\alpha^t(\tau) + \left(1 - \alpha^t(\tau)\right)\left(1 - \alpha^t(\tau)\right)\Pr(\text{two}|\tau),$$

so that

$$\Pr(\text{two}|\tau) = \frac{\alpha^t(\tau)}{2 - \alpha^t(\tau)}. \tag{7.31}$$

Thus firm i invests while firm j does not invest with probability $\frac{1-\alpha^t(t)}{2-\alpha^t(t)}$, with the same probability firm j invests while firm i does not invest, and with probability $\frac{\alpha^t(t)}{2-\alpha^t(t)}$ both firms invest at the same time. Consequently, if $\alpha^t(\tau) = 0$ we have

$$\Pr(\text{one}|\tau) = \frac{1}{2}, \tag{7.32}$$

$$\Pr(\text{two}|\tau) = 0. \tag{7.33}$$

FIRST CASE: PREEMPTION

For the moment assume that one firm, say firm i, has been given the leader role beforehand, thus firm j can only decide to invest after firm i has done so. The optimal investment time for the leader in the first case, thus where expression (7.28) holds, is denoted by

$$T_L = \inf\left(t\,|\,Y(t) \geq Y_L\right), \tag{7.34}$$

where

$$Y_L = \frac{\beta_1}{\beta_1 - 1}\frac{(r - \mu)I}{D_{10} - D_{00}}. \tag{7.35}$$

The threshold Y_L is derived by solving the value matching and smooth pasting conditions that result from the leader's optimal stopping problem (see Appendix A.2). The uniqueness of the threshold can be proved using the same steps as in the proof of Proposition 7.1. Note that Y_L is equal to Y_M. The reason is that for $Y \in (0, Y_F)$ the leader's decision has no effect on the optimal reply of the follower. Therefore the leader acts as if there is no follower, and thus behaves like a monopolist. As D_{10} increases it is more attractive to be the first investor so that Y_L, and thus the expected value of T_L, decreases. In Appendix B we prove the following proposition.

PROPOSITION 7.2 *It holds that*

$$L(Y_L) > F(Y_L). \tag{7.36}$$

Now let us drop the assumption that one firm is given the leader role beforehand. Then the implication of Proposition 7.2 is that each firm wants to be the only one to invest at time T_L. A firm will try to preempt its competitor by investing at time $T_L - \varepsilon$, since it knows that the other firm would like to be the first to invest at time T_L. But then the other firm will try to invest at time $T_L - 2\varepsilon$. This process of preemption stops at time T_P, where

$$T_P = \inf(t | Y(t) \geq Y_P), \tag{7.37}$$

in which Y_P is the solution of the following equation

$$L(Y_P) = F(Y_P).$$

Before time T_P there are no incentives to become leader, since for $t < T_P$ the follower payoff exceeds the leader payoff. This is because $t < T_P$ implies that $Y < Y_P$, which in turn implies that $F(Y) > L(Y)$ due to the fact that Y_P is unique as stated in the following proposition. The proof can be found in Appendix B.

PROPOSITION 7.3 *There exists a unique value for Y, Y_P, such that*

$$L(Y_P) = F(Y_P) \ and \ 0 < Y_P < Y_F. \tag{7.38}$$

For this first case the payoff curves are depicted in Figure 7.2. The investment opportunity is worthless for Y equal to 0. Therefore at $Y = 0$ the leader (L) and joint investment (M) value equal minus the investment cost and the follower (F) and optimal joint investment value (J) equal zero. M is a linear increasing function of Y (see equation (7.22)). The follower has the choice between investing at the same time as the leader or to wait. Since the optimal follower action on the interval $(0, Y_F)$ is to wait, the follower curve is situated above the joint investment curve on that interval. From Subsection 4.2 we know that the leader, follower and joint investment curves coincide with each other for Y larger than or equal to Y_F. Due to the existence and uniqueness of Y_P (see Proposition 7.3), the leader curve crosses the follower curve once on the interval $(0, Y_F)$ (at Y_P). Since (7.28) holds here, the leader curve also crosses the optimal joint investment curve somewhere on the interval $(0, Y_F)$. For Y larger or equal than Y_J the joint investment curve coincides with the other three curves.

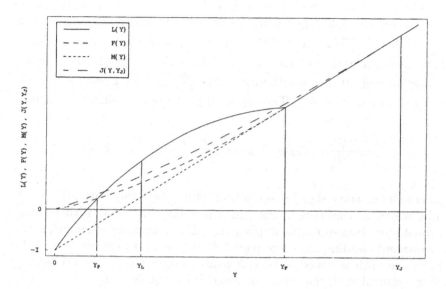

Figure 7.2. First Case: Preemption

The equilibrium strategy of firm $i \in \{1, 2\}$ equals (cf. Appendix A.2)

$$G_i^t(s) \;=\; G(s) = \begin{cases} 0 & \text{if } s < T_P, \\ 1 & \text{if } s \geq T_P, \end{cases} \tag{7.39}$$

$$\alpha_i^t(s) \;=\; \alpha(s) = \begin{cases} 0 & \text{if } s < T_P, \\ \dfrac{L(Y(s)) - F(Y(s))}{L(Y(s)) - M(Y(s))} & \text{if } T_P \leq s < T_F, \\ 1 & \text{if } s \geq T_F. \end{cases} \tag{7.40}$$

The equilibrium outcome depends on the value $y\,(= Y(0))$. Three regions have to be distinguished.

The first region is defined by $y \leq Y_P$. There are two possible equilibrium outcomes. In the first outcome firm 1 is the leader and invests at time T_P and firm 2 is the follower and invests at time T_F. The second outcome is the symmetric counterpart: firm 2 is the leader and invests at time T_P and firm 1 is the follower and invests at time T_F. Since at time T_P it holds that $Y = Y_P$, it can be obtained from (7.38) and (7.40) that $\alpha(T_P) = 0$. Due to (7.32) it can be concluded that each outcome occurs with probability one-half. Furthermore, from (7.33) we get that the probability that both firms invest simultaneously is zero. Due to (7.38), it follows that the expected value of each firm equals $F(Y_P)$.

In the second region it holds that $Y_P < y < Y_F$. There are three possible outcomes. Since L exceeds F in case $Y \in (Y_P, Y_F)$, it can be obtained from (7.40) that $\alpha(0) > 0$. Due to (7.30) we know that with probability $\frac{1-\alpha(0)}{2-\alpha(0)}$ anyone of the firms invests at time 0 and the other firm invests at time T_F. Expression (7.31) implies that the firms invest simultaneously at time 0 with probability $\frac{\alpha(0)}{2-\alpha(0)}$, leaving them with a low value of $M(y)\,(< F(y))$. The expected payoff of each firm thus equals

$$\frac{1-\alpha(0)}{2-\alpha(0)}\left(L(y) + F(y)\right) + \frac{\alpha(0)}{2-\alpha(0)}M(y) = F(y),$$

where the equality sign follows from (7.40). Since there are first mover advantages in this region, each firm is willing to invest with positive probability. However, this implies that the probability of simultaneous investment, leading to a low payoff $M(y)$, is also positive. Since the firms are both assumed to be risk neutral, they will fix the probability of investment such that their expected value equals $F(y)$, which is also their payoff if they let the other firm invest first.

When y is in the third region $[Y_F, \infty)$, the outcome exhibits joint investment at time 0. The expected value of each firm is equal to $M(y) = L(y) = F(y)$.

SECOND CASE: JOINT INVESTMENT

In the case of joint investment expression (7.29) holds, which leads to Figure 7.3. There turn out to be an infinite number of symmetric equilibrium strategies, which can be divided into two classes. The first class consists of the strategy described above (see equations (7.39)-(7.40)). The second class consists of strategies where firms invest simultaneously. They have the following form $(i \in \{1,2\})$:

$$G_i^t(s) = G(s) = \begin{cases} 0 & \text{if } s < T^*, \\ 1 & \text{if } s \geq T^*, \end{cases} \tag{7.41}$$

$$\alpha_i^t(s) = \alpha(s) = \begin{cases} 0 & \text{if } s < T^*, \\ 1 & \text{if } s \geq T^*, \end{cases} \tag{7.42}$$

for any $T^* \in [T_S, T_J]$ where

$$T_S = \inf(t \mid Y(t) \geq Y_S),$$
$$Y_S = \min(\theta \mid J(Y, \theta) \geq L(Y) \text{ for all } Y \geq 0).$$

In Figure 7.4 the construction of Y_S is shown graphically. In that figure the functions $J(Y, Y_J) - L(Y)$ and $J(Y, Y_S) - L(Y)$ are plotted. To find Y_S one starts out with the function $J(Y, \theta) - L(Y)$ with $\theta = Y_J$. Then θ is lowered. The lowest θ for which the function is still non-negative is Y_S. Note that the function $J(Y, Y_S) - L(Y)$ has exactly one point of tangency on the interval $(0, Y_F)$. Thus the curves $J(Y, Y_S)$ and $L(Y)$ meet tangent at that point, which is Y_L. This result can be shown mathematically by solving the following two equations simultaneously for $Y (\in (0, Y_F))$: $J(Y, \theta) = L(Y)$ and $\frac{\partial J(Y, \theta)}{\partial Y} = \frac{\partial L(Y)}{\partial Y}$.

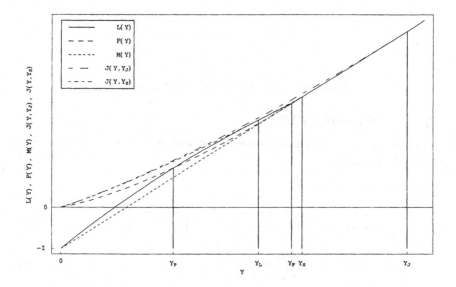

Figure 7.3.   Second Case: Collusion

From (7.26) it can be concluded that the equilibrium that is supported by the strategies (7.41)-(7.42) with $T^* = T_J$ is the Pareto dominant equilibrium and therefore the most reasonable outcome in this case. In what follows we assume that the Pareto dominant equilibrium is indeed the outcome in the second case. This would have been the only equilibrium if we apply the setup described in Simon, 1987a. For this equilibrium it holds that there is simultaneous investment at time T_J. The expected value of each firm equals $J(y, Y(T_J))$.

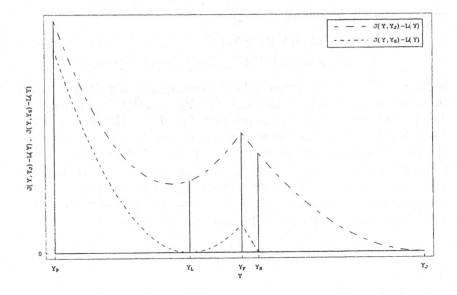

Figure 7.4.   Construction of $J(Y, Y_S)$ curve.

4.5   PROPERTIES

The following proposition states when which case applies. See Appendix B for a proof.

PROPOSITION 7.4 *Define*

$$f(\beta_1) = \beta_1 \left(\frac{D_{10} - D_{11}}{D_{11} - D_{01}} \right) + \left(\frac{D_{11} - D_{00}}{D_{11} - D_{01}} \right)^{\beta_1}, \qquad (7.43)$$

$$g(\beta_1) = \left(\frac{D_{10} - D_{00}}{D_{11} - D_{01}} \right)^{\beta_1}. \qquad (7.44)$$

Whenever the following inequality holds the equilibrium is of the preemption type and otherwise of the joint investment type:

$$f(\beta_1) < g(\beta_1). \qquad (7.45)$$

Proposition 7.4 implies that the equilibrium is always of the preemption type, no matter the value of β_1 and thus the degree of uncertainty, if D_{10} is large enough, i.e. if the incentives to become leader are large enough. If D_{10} is relatively small, the incentives to become leader almost vanish and the joint investment equilibrium turns up.

Note that condition (7.45) is independent of the value of the investment cost I (as long as it is strictly positive). This for the reason that changing the investment cost only changes the absolute values of the investment triggers, and therefore the value functions, but not the relative values.

The following proposition, that is proved in Appendix B, states the effect of β_1 on the type of the equilibrium.

PROPOSITION 7.5 *There are three different scenarios:*

(i) *If $f'(1) > g'(1)$ and $f\left(\frac{r}{\mu}\right) \geq g\left(\frac{r}{\mu}\right)$ the equilibrium is always of the joint investment type.*

(ii) *If $f'(1) > g'(1)$ and $f\left(\frac{r}{\mu}\right) < g\left(\frac{r}{\mu}\right)$ the equilibrium is of the joint investment type for relatively low values of β_1 and of the preemption type for relatively high values of β_1.*

(iii) *If $g'(1) \geq f'(1)$ the equilibrium is always of the preemption type.*

Propositions 7.4 and 7.5 are visualized in Figure 7.5. In that figure we have plotted the function $g(\beta_1)$, the boundary between the preemption case and joint investment case (Proposition 7.4), and for each possible scenario the corresponding $f(\beta_1)$ function (Proposition 7.5).

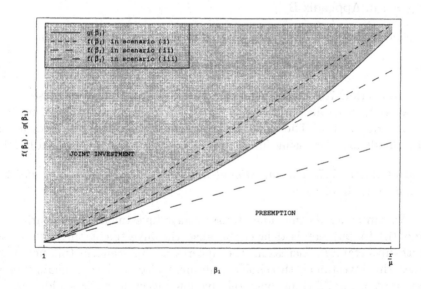

Figure 7.5. Possible scenarios.

In scenario (i) the first mover advantage is that large that the preemption equilibrium will always turn up. The opposite is going on in scenario (ii), where the first mover advantage is that low that the equilibrium where both firms invest jointly at a later point in time is the Pareto-dominant equilibrium.

Hence, only in scenario (iii) the type of equilibrium depends on β_1. The economic implications are stated in the following corollary to Proposition 7.5. The proof can be found in Appendix B.

COROLLARY 7.1 *In scenario (iii) the equilibrium is of the joint investment (preemption) type for high (low) values of σ and μ and low (high) values of r.*

Here, it is natural that the preemption equilibrium arises if there is not much uncertainty (low σ), the μ is low, or the interest rate is high, since then the value of waiting is low. The contrary holds in very uncertain (high σ) economic environments, or environments where the μ is high, or the interest rate is low. Then the value of waiting is large, which implies that investing faces high opportunity costs. This makes a preemption strategy unattractive.

Proposition 7.6 compares the investment thresholds of the duopoly model with the investment threshold of the monopoly model. The proof is given in Appendix B.

PROPOSITION 7.6 *For every parameter configuration it holds that*

$$Y_P \leq Y_M < Y_J. \tag{7.46}$$

Proposition 7.6 implies that the speed of investment increases (decreases) if strategic interactions result in a preemption (joint investment) equilibrium. The following proposition states that the investment thresholds are decreasing functions of β_1 (for a proof see Appendix B).

PROPOSITION 7.7 *The investment thresholds Y_P, Y_L, Y_M, Y_F, and Y_J are decreasing in β_1.*

We can conclude that uncertainty delays investment. In scenarios (i) and (ii) investment is delayed because the investment thresholds rise with uncertainty. Increasing the uncertainty in scenario (iii) not only rises the investment thresholds, but may also lead to a change of a preemption equilibrium (with relative low investment thresholds) into a joint investment equilibrium (with relative high investment thresholds).

In the real options literature it is argued that an investment should be undertaken when the net present value exceeds the option value of waiting. For models with strategic interactions this investment rule should be changed: investing is optimal when the net present value exceeds the *strategic option value of waiting.* The strategic option value of waiting incorporates the money value of the strategic interactions in the option value of waiting. The following proposition compares the strategic option value of waiting in the duopoly case with the option value of waiting in the monopoly case. The proof is given in Appendix B.

PROPOSITION 7.8 *Compared to the option value of waiting in the monopoly case (at Y_M, see (7.14)),*

$$\frac{I}{\beta_1 - 1},$$

the strategic option value of waiting is

1. *smaller in the preemption case (at Y_P);*

2. *the same in the joint investment case (at Y_J).*

When strategic interactions lead to a preemption equilibrium, it is even possible that the firms make an investment with a negative net present value. Then the strategic option value of waiting is negative. For example, take the following parameter values $D_{10} = 10$, $D_{11} = 4$, $D_{00} = 2$, $D_{01} = 1$, $r = 0.10$, $\mu = 0.05$, and $I = 10$. For these parameters equation (7.45) is always satisfied so that the equilibrium is always of the preemption type. The net present value of investment at Y_P equals $L(Y_P) - \frac{Y_P D_{00}}{r - \mu}$. In the left part of Figure 7.6 this net present value is plotted as function of σ. For sake of comparison, in the right part the corresponding net present value of investment for the monopolist is presented.

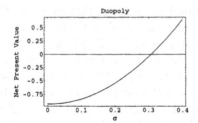 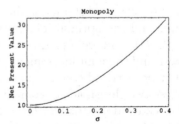

Figure 7.6. Net present value of investment of duopolist at Y_P and monopolist at Y_M as function of σ.

As in the monopoly case, the option value of waiting still increases with uncertainty in the duopoly model. From Figure 7.6 we conclude

178 TECHNOLOGY INVESTMENT

that for low (but realistic) values of σ ($\sigma < 0.308$) strategic interactions lead to a negative strategic option value of waiting. Thus the strategic interactions force the firms to make an investment with a negative net present value. Becoming inactive is even worse for the firms, since then the net present value would equal $-\frac{Y_P D_{00}}{r-\mu}$.

5.   EXISTING LITERATURE

In this section we confront our results with the existing literature. Our model is an extension of the model described in Chapter 9 of Dixit and Pindyck, 1996. Contrary to that model we also allow that before the moment of investment the firms are already active on the output market on which they compete. Nielsen, 1999 showed that in the Dixit and Pindyck, 1996 model competition on the output market decreases the option value of waiting and therefore duopolistic firms will invest earlier than monopolistic firms. Remember that for $D_{00} = D_{01} = 0$ our model is the Dixit and Pindyck, 1996 model. Since for $x = \frac{D_{10}}{D_{11}} > 1$ we have that $\beta_1 x - (\beta_1 - 1) < x^{\beta_1}$, equation (7.46) is always satisfied, which implies that the equilibrium is always of the preemption type in the Dixit and Pindyck, 1996 model. Thus Nielsen, 1999's result is a consequence of the initial conditions on the output market in the sense that if both firms are initially active on the output market, Nielsen, 1999's result does not hold anymore in general. From Proposition 7.6 we conclude that there are two possibilities. In the case where a joint investment equilibrium is the most reasonable outcome, strategic interactions result in delayment of investment by the firms. In the case where the only equilibrium is of the preemption type, competition accelerates investment if we compare the moment of investment of the leader in the duopoly to the monopolist.

In the new market model the optimal investment threshold for the follower and the optimal joint investment threshold coincide (cf. (7.18) and (7.27)). Therefore, due to (7.16) and (7.25) it follows that the follower and the joint investment curve coincide, which implies that there can not be a second case in the new market model. The economic reason for these two thresholds to coincide is the fact that the investment timing of the leader does not affect the follower's profit flow in a new market model, whereas in our model the follower's profit flow decreases from $Y D_{00}$ to $Y D_{01}$ at the moment the leader invests. Thus, in our model the follower will invest earlier to recapture market share from the leader. That is why we have $Y_F < Y_J$.

At present, only a few contributions deal with the effect of strategic interactions on the option value of waiting associated with investments under uncertainty. However, in these papers the coordination problem is avoided, and thus not treated in the way we did in Section 4. For instance, Weeds, 1999 implicitly makes the unsatisfactory assumption that only one firm will succeed in investing in case there is an incentive to be the first firm to invest and it is only optimal for one firm to invest. There are two reasons for that assumption to be unsatisfactory: (1) the firms are imposed to be equal and (2) the firms can invest simultaneously if it is optimal for both. Note that this assumption, although explicitly, is also made in Nielsen, 1999, Grenadier, 1996, and Dutta et al., 1995.

The reason for our outcomes to be more realistic is as follows. When there is an incentive to be the first to invest ($L > F > M$) both firms are willing to take a risk and since they are both assumed to be risk neutral they will risk so much that their expected value equals F, which equals their payoff if they allow the other firm to invest first. Employing the results of Section 4 learns that in this case both firms set $\alpha = \frac{L-F}{L-M}$, and that there is a positive probability $\frac{\alpha}{2-\alpha}$ that both firms invest exactly at the same time, leaving them with the low payoff M.

On p. 313 of Dixit and Pindyck, 1996 it is claimed that in the Smets, 1991 model, the probability that both firms invest simultaneously, while it is only optimal for one firm to invest, is always zero. From the above argumentation it should be clear by now that this claim is not correct. To correct another point, consider page 314 of Dixit and Pindyck, 1996. First, note that their threshold Y_2 is equal to our threshold Y_F and their threshold Y_3 equals our threshold Y_J. Now, we know that in the new market model we have that $Y_J = Y_F$, so that Y_3 is equal to Y_2 in their model and not, as they claim, greater than Y_2.

6. CONCLUSIONS

This chapter brings together two streams of literature: investment under competition (Reinganum, 1981, Fudenberg and Tirole, 1985, and Chapter 4) and investment under uncertainty (Dixit and Pindyck, 1996). In this concluding section we focus on the question how introduction of uncertainty changes the results derived for the deterministic duopoly framework of Fudenberg and Tirole, 1985. In Fudenberg and Tirole it was obtained that under large first mover advantages a preemption equilibrium with dispersed adoption timings results, while otherwise a joint adoption equilibrium is the Pareto-dominant outcome.

After introduction of uncertainty the firm's investment timing problem has to deal with the option value of waiting: when a firm makes an irreversible investment expenditure, it exercises its option to invest. It gives up the possibility of waiting for new information to arrive that might affect the desirability or timing of the expenditure. It is clear that a huge option value of waiting, which arises in highly uncertain economic environments, results in a considerable delayment of investment. On the other hand, in the preemption equilibrium of Fudenberg and Tirole, 1985 it is imperative for a firm to invest quickly and thereby preempt investment by potential competitors.

Our chapter brings these contrary forces together and it turns out that our results relate to those of Fudenberg and Tirole, 1985 in the following way. Whenever Fudenberg and Tirole conclude that joint adoption is the Pareto-dominant outcome, this also holds for our model. Also, if first mover advantages are sufficiently large, for both models the preemption equilibrium results, but in the stochastic case for both firms the investment timing is delayed by the option value of waiting. Finally, if first mover advantages are a bit lower, but still high enough for the preemption equilibrium to prevail in the deterministic framework of Fudenberg and Tirole, 1985, introduction of sufficiently large uncertainty results in a joint adoption equilibrium that Pareto-dominates all other equilibria. This brings us to the conclusion that introduction of uncertainty reduces the number of scenarios under which the preemption equilibrium is the optimal outcome.

Appendices
A.   DERIVATION OF VALUE FUNCTIONS
A.1   MONOPOLY

Let $Y(t)$ behave according to equations (7.3) and (7.4). Then Itô's lemma (see Appendix A of Chapter 2) implies that

$$d \log Y(t) = \left(\mu - \tfrac{1}{2}\sigma^2\right) dt + \sigma d\omega(t).$$

Therefore

$$\log Y(t) = \log y + \int_{s=0}^{t} \left(\mu - \tfrac{1}{2}\sigma^2\right) ds + \int_{s=0}^{t} \sigma d\omega(s).$$

Thus for $t \geq 0$ it holds that

$$Y(t) = y e^{\left(\mu - \frac{1}{2}\sigma^2\right)t + \sigma\omega(t)}, \tag{7.47}$$

where $\omega(t)$ is distributed according to a normal distribution with mean zero and variance t.

In the stopping region the value of the firm equals

$$V\left(Y\left(t\right)\right) \;=\; E\left[\int_{\tau=t}^{\infty} Y\left(\tau\right) D_{10} e^{-r(\tau-t)} d\tau\right] - I$$

$$=\; E\left[\int_{\tau=0}^{\infty} Y\left(t\right) e^{\left(\mu-\frac{1}{2}\sigma^2\right)\tau+\sigma w(\tau)} D_{10} e^{-r\tau} d\tau\right] - I$$

$$=\; Y\left(t\right) D_{10} \int_{\tau=0}^{\infty}\int_{w=-\infty}^{\infty} \frac{1}{\sqrt{\tau}\sqrt{2\pi}} e^{-\frac{w^2}{2\tau}-\left(r-\mu+\frac{1}{2}\sigma^2\right)\tau+\sigma w} dw d\tau - I$$

$$=\; Y\left(t\right) D_{10} \int_{\tau=0}^{\infty} e^{-\left(r-\mu+\frac{1}{2}\sigma^2\right)\tau}$$

$$\times\; \int_{w=-\infty}^{\infty} \frac{1}{\sqrt{\tau}\sqrt{2\pi}} e^{-\frac{(w-\sigma\tau)^2}{2\tau}+\frac{1}{2}\sigma^2\tau} dw d\tau - I$$

$$=\; Y\left(t\right) D_{10} \int_{\tau=0}^{\infty} e^{-\left(r-\mu+\frac{1}{2}\sigma^2\right)\tau} e^{\frac{1}{2}\sigma^2\tau} d\tau - I$$

$$=\; \frac{Y\left(t\right) D_{10}}{r-\mu} - I.$$

The value function must satisfy the following Bellman equation in the continuation region

$$rV\left(Y\right) = YD_{00} + \lim_{dt\downarrow 0}\frac{1}{dt}E\left[dV\left(Y\right)\right]. \tag{7.48}$$

Applying Itô's Lemma (see Appendix A of Chapter 2) to the expectation in the right-hand side of equation (7.48) gives

$$E\left[dV\left(Y\right)\right] = \left(\mu Y\frac{\partial V\left(Y\right)}{\partial Y} + \frac{1}{2}\sigma^2 Y^2\frac{\partial^2 V\left(Y\right)}{\partial Y^2}\right) dt + o\left(dt\right). \tag{7.49}$$

Substitution of equation (7.49) into equation (7.48) and rewriting leads to the following differential equation

$$-rV\left(Y\right) + \mu Y\frac{\partial V\left(Y\right)}{\partial Y} + \frac{1}{2}\sigma^2 Y^2\frac{\partial^2 V\left(Y\right)}{\partial Y^2} + YD_{00} = 0. \tag{7.50}$$

The general solution of (7.50) is given by

$$V\left(Y\right) = A_1 Y^{\beta_1} + A_2 Y^{\beta_2} + \frac{YD_{00}}{r-\mu}, \tag{7.51}$$

where β_1 (β_2) is the positive (negative) root of the following quadratic equation

$$\frac{1}{2}\sigma^2\beta^2 + \left(\mu - \frac{1}{2}\sigma^2\right)\beta - r = 0. \tag{7.52}$$

When $Y = 0$ the value of the firm is equal to zero so that the boundary condition $V(0) = 0$ leads to $A_2 = 0$.

A.2   LEADER

Given that $Y < Y_F$ the value of the leader must satisfy the following Bellman equation

$$r\left(L(Y) + I\right) = YD_{10} + \lim_{dt\downarrow 0}\frac{1}{dt}E\left[dL(Y)\right]. \tag{7.53}$$

Itô's Lemma gives

$$r\left(L(Y) + I\right) = YD_{10} + \mu Y\frac{\partial L(Y)}{\partial Y} + \frac{1}{2}\sigma^2 Y^2\frac{\partial^2 L(Y)}{\partial Y^2}. \tag{7.54}$$

The general solution of (7.54) is equal to

$$L(Y) = E_1 Y^{\beta_1} + E_2 Y^{\beta_2} + \frac{YD_{10}}{r - \mu} - I. \tag{7.55}$$

The following two boundary conditions should be satisfied

$$L(0) = 0, \tag{7.56}$$

$$L(Y_F) = \frac{Y_F D_{11}}{r - \mu} - I. \tag{7.57}$$

Equations (7.56) and (7.57) lead to

$$E_1 = Y_F^{1-\beta_1}\frac{D_{11} - D_{10}}{r - \mu}, \tag{7.58}$$

$$E_2 = 0. \tag{7.59}$$

Before the leader has invested its value equals (same derivation as in Appendix A.1)

$$K_1 Y^{\beta_1} + \frac{YD_{00}}{r - \mu}. \tag{7.60}$$

Next let us derive an expression for Y_L. The following value matching and smooth pasting conditions must be satisfied

$$K_1 Y_L^{\beta_1} + \frac{Y_L D_{00}}{r - \mu} = E_1 Y_L^{\beta_1} + \frac{Y_L D_{10}}{r - \mu} - I, \tag{7.61}$$

$$\beta_1 K_1 Y_L^{\beta_1 - 1} + \frac{D_{00}}{r - \mu} = \beta_1 E_1 Y_L^{\beta_1 - 1} + \frac{D_{10}}{r - \mu}. \tag{7.62}$$

Solving these last two equations yields

$$Y_L = \frac{\beta_1}{\beta_1 - 1} \frac{(r - \mu) I}{D_{10} - D_{00}}, \tag{7.63}$$

$$K_1 = E_1 + \frac{Y_L^{1-\beta_1}}{\beta_1} \frac{D_{10} - D_{00}}{r - \mu}. \tag{7.64}$$

B.   LEMMAS AND PROOFS

PROOF OF PROPOSITION 7.1 *Theorem 2.4 gives sufficient conditions for the uniqueness of the threshold. Here, the functions π and Ω are given by*

$$\pi(Y) = Y D_{00},$$
$$\Omega(Y) = \frac{Y D_{10}}{r - \mu} - I.$$

The function

$$\pi(Y) - r\Omega(Y) + \lim_{dt \downarrow 0} \frac{1}{dt} E\left[d\Omega(Y) | Y\right]$$

$$= \pi(Y) - r\Omega(Y) + \mu Y \frac{\partial \Omega(Y)}{\partial Y} + \frac{1}{2}\sigma^2 Y^2 \frac{\partial^2 \Omega(Y)}{\partial Y^2}$$

$$= Y D_{00} - r\left(\frac{Y D_{10}}{r - \mu} - I\right) + \mu Y \frac{D_{10}}{r - \mu}$$

$$= Y (D_{00} - D_{10}) + rI,$$

is indeed decreasing in Y since $D_{00} < D_{10}$.

The positive persistence of uncertainty property is also satisfied. It holds that

$$\Phi(y|x) = \Pr(Y(t + dt) \leq y | Y(t) = x)$$
$$= \Pr(Y(t) + dY(t) \leq y | Y(t) = x)$$
$$= \Pr(dY(t) \leq y - x | Y(t) = x)$$
$$= \Pr(\mu x dt + \sigma x d\omega(t) \leq y - x)$$
$$= \Pr\left(d\omega(t) \leq \frac{y - x - \mu x dt}{\sigma x}\right)$$
$$= \int_{z=-\infty}^{\frac{y-x-\mu x dt}{\sigma x}} \frac{1}{\sqrt{2\pi}} e^{-\frac{1}{2}z^2} dz.$$

Thus for $x_1 < x_2$ we have that $\Phi(y|x_1) > \Phi(y|x_2)$, because

$$\frac{\partial \frac{y-x-\mu x dt}{\sigma x}}{\partial x} = -\frac{y}{\sigma x^2} < 0.$$

□

PROOF OF PROPOSITION 7.2 *Define the function ϕ as follows*

$$\phi(Y) = L(Y) - F(Y). \tag{7.65}$$

Then we have to prove that

$$\phi(Y_L) > 0. \tag{7.66}$$

For $Y \in [0, Y_F]$ the value of the follower, in case the leader has already invested, can be expressed by

$$F(Y) = \frac{Y D_{01}}{r - \mu} + \left(\frac{Y}{Y_F}\right)^{\beta_1} \left(\frac{Y_F(D_{11} - D_{01})}{r - \mu} - I\right).$$

For $Y \in [0, Y_F]$ the function $\phi(Y)$ equals

$$
\begin{aligned}
\phi(Y) &= L(Y) - F(Y) \\
&= \frac{Y D_{10}}{r - \mu} - I + \left(\frac{Y}{Y_F}\right)^{\beta_1} \frac{Y_F(D_{11} - D_{10})}{r - \mu} \\
&\quad - \frac{Y D_{01}}{r - \mu} - \left(\frac{Y}{Y_F}\right)^{\beta_1} \left(\frac{Y_F(D_{11} - D_{01})}{r - \mu} - I\right) \\
&= \frac{Y(D_{10} - D_{01})}{r - \mu} - I - \left(\frac{Y}{Y_F}\right)^{\beta_1} \left(\frac{Y_F(D_{10} - D_{01})}{r - \mu} - I\right). \tag{7.67}
\end{aligned}
$$

Substitution of equation (7.35) into (7.67) gives

$$\phi(Y_L) = I \left[\frac{\beta_1}{\beta_1 - 1} \frac{D_{10} - D_{01}}{D_{10} - D_{00}} - 1 - \left(\frac{D_{11} - D_{01}}{D_{10} - D_{00}}\right)^{\beta_1} \left(\frac{\beta_1}{\beta_1 - 1} \frac{D_{10} - D_{01}}{D_{11} - D_{01}} - 1\right)\right]. \tag{7.68}$$

Define

$$a = \frac{D_{10} - D_{01}}{D_{10} - D_{00}} > 1, \tag{7.69}$$

$$b = \frac{D_{11} - D_{01}}{D_{10} - D_{00}} < 1. \tag{7.70}$$

The inequalities hold due to equations (7.5) and (7.6). After substitution of (7.69) and (7.70) into (7.68) it is obtained that

$$h(a, b) = \frac{(\beta_1 - 1)\,\phi(Y_L)}{I} = \beta_1 a - (\beta_1 - 1) - \beta_1 a b^{\beta_1 - 1} + (\beta_1 - 1) b^{\beta_1}. \tag{7.71}$$

The proposition is proved if we show that $h(a, b) > 0$ for all $a \in (1, \infty)$ and $b \in (0, 1)$. This holds since

$$\frac{\partial h(a, b)}{\partial a} = \beta_1 - \beta_1 b^{\beta_1 - 1} > 0, \tag{7.72}$$

$$\frac{\partial h(a, b)}{\partial b} = -(\beta_1 - 1)\beta_1 a b^{\beta_1 - 2} + \beta_1(\beta_1 - 1)b^{\beta_1} < 0, \tag{7.73}$$

$$h(1, 1) = 0. \tag{7.74}$$

□

LEMMA 7.1 *Define the function ϕ as follows*

$$\phi(Y) = L(Y) - F(Y). \tag{7.75}$$

Then it holds that

$$\phi(0) < 0, \tag{7.76}$$

$$\phi(Y_F) = 0, \tag{7.77}$$

$$\left. \frac{\partial \phi(Y)}{\partial Y} \right|_{Y=Y_F} < 0, \tag{7.78}$$

$$\frac{\partial^2 \phi(Y)}{\partial Y^2} \leq 0 \text{ for all } Y \geq 0. \tag{7.79}$$

PROOF OF LEMMA 7.1 *For $Y \in [0, Y_F]$ the function $\phi(Y)$ equals (see (7.67)):*

$$\phi(Y) = \frac{Y(D_{10} - D_{01})}{r - \mu} - I - \left(\frac{Y}{Y_F}\right)^{\beta_1} \left(\frac{Y_F(D_{10} - D_{01})}{r - \mu} - I\right). \tag{7.80}$$

Expressions (7.76) and (7.77) follow directly after setting $Y = 0$ and $Y = Y_F$, respectively, in equation (7.80).

The first derivative of $\phi(Y)$ equals

$$\frac{\partial \phi(Y)}{\partial Y} = \frac{D_{10} - D_{01}}{r - \mu} - \beta_1 \frac{Y^{\beta_1 - 1}}{Y_F^{\beta_1}} \left(\frac{Y_F(D_{10} - D_{01})}{r - \mu} - I\right). \tag{7.81}$$

Setting $Y = Y_F$ in equation (7.81) gives

$$\left. \frac{\partial \phi(Y)}{\partial Y} \right|_{Y=Y_F} = \frac{D_{10} - D_{01}}{r - \mu} - \beta_1 \frac{1}{Y_F} \left(\frac{Y_F(D_{10} - D_{01})}{r - \mu} - I\right). \tag{7.82}$$

Substitution of equation (7.18) and rearranging gives

$$\left. \frac{\partial \phi(Y)}{\partial Y} \right|_{Y=Y_F} = -(\beta_1 - 1)\left(\frac{D_{10} - D_{11}}{r - \mu}\right) < 0, \tag{7.83}$$

which confirms (7.78). The second derivative of $\phi(Y)$ is given by

$$\frac{\partial^2 \phi(Y)}{\partial Y^2} = -\beta_1(\beta_1 - 1)\frac{Y^{\beta_2 - 1}}{Y_F^{\beta_1}}\left(\frac{Y_F(D_{10} - D_{01})}{r - \mu} - I\right)$$

$$= -\beta_1(\beta_1 - 1)\frac{Y^{\beta_2 - 1}}{Y_F^{\beta_1}}\left(\left(\frac{\beta_1}{\beta_1 - 1}\right)\left(\frac{D_{10} - D_{01}}{D_{11} - D_{01}}\right) - 1\right)I. \quad (7.84)$$

Expression (7.79) follows from equation (7.84) since $\beta_1 > 1$ and $D_{10} > D_{11}$. □

PROOF OF PROPOSITION 7.3 *Proposition 7.3 is a direct result of Lemma 7.1.* □

LEMMA 7.2 *For $0 < Y \leq Y_F$ it holds that*

$$J(Y, Y_J) > F(Y). \quad (7.85)$$

PROOF OF LEMMA 7.2 *It is obvious that $J(Y, Y_F) > F(Y)$ for $Y \in (0, Y_F)$, since $D_{00} > D_{01}$. And by definition it holds that $J(Y, Y_J) \geq J(Y, Y_F)$.* □

LEMMA 7.3 *Define the function $\gamma(Y)$ as follows*

$$\gamma(Y) = J(Y, Y_J) - L(Y). \quad (7.86)$$

Then the following properties hold:

$$\gamma(Y_P) > 0, \quad (7.87)$$
$$\gamma(Y_F) > 0, \quad (7.88)$$
$$\frac{\partial^2 \gamma(Y)}{\partial Y^2} > 0. \quad (7.89)$$

PROOF OF LEMMA 7.3 *Substitution of equations (7.21) and (7.25) into equation (7.86) gives*

$$\gamma(Y) = \frac{Y(D_{00} - D_{10})}{r - \mu} + I + (H_1 - E_1)Y^{\beta_1},$$

where $E_1 (< 0)$ is given by equation (7.58) and

$$H_1 = \frac{Y_J^{1-\beta_1}}{\beta_1}\left(\frac{D_{11} - D_{00}}{r - \mu}\right) > 0. \quad (7.90)$$

Properties (7.87) and (7.88) follow from Lemma 7.2 together with

$$L(Y_P) = F(Y_P), \quad (7.91)$$
$$L(Y_F) = F(Y_F). \quad (7.92)$$

The second derivative of $\gamma(Y)$ is equal to

$$\frac{\partial^2 \gamma(Y)}{\partial Y^2} = (\beta_1 - 1)\beta_1 (H_1 - E_1) Y^{\beta_1 - 2}. \tag{7.93}$$

Remembering that $E_1 < 0$ and $H_1 > 0$ gives equation (7.89). □

LEMMA 7.4 *It holds that*

$$\min_{Y \geq 0} \gamma(Y) < 0, \tag{7.94}$$

if and only if

$$f(\beta_1) < g(\beta_1). \tag{7.95}$$

PROOF OF LEMMA 7.4 *The first derivative of $\gamma(Y)$ is given by*

$$\frac{\partial \gamma(Y)}{\partial Y} = \frac{D_{00} - D_{10}}{r - \mu} + \beta_1 (H_1(Y_J) - E_1) Y^{\beta_1 - 1}. \tag{7.96}$$

The solution of

$$\frac{\partial \gamma(Y)}{\partial Y} = 0, \tag{7.97}$$

equals

$$Y^* = \left(\frac{D_{10} - D_{00}}{\beta_1 (H_1 - E_1)(r - \mu)} \right)^{\frac{1}{\beta_1 - 1}} > 0. \tag{7.98}$$

The minimum (expression (7.88) implies that $\gamma(Y^*)$ is a unique minimum) of γ equals

$$\gamma(Y^*) = I + (D_{10} - D_{00})^{\frac{\beta_1}{\beta_1 - 1}} (H_1 - E_1)^{\frac{1}{1 - \beta_1}} (r - \mu)^{\frac{\beta_1}{1 - \beta_1}}$$
$$\times \left(\beta_1^{\frac{\beta_1}{1 - \beta_1}} - \beta_1^{\frac{1}{1 - \beta_1}} \right). \tag{7.99}$$

The minimum is negative if and only if (substitute (7.58) and (7.90) in (7.99) and rewrite)

$$\left[\left(\frac{D_{11} - D_{00}}{D_{10} - D_{00}} \right)^{\beta_1} + \beta_1 \left(\frac{D_{10} - D_{11}}{D_{11} - D_{01}} \right) \left(\frac{D_{11} - D_{01}}{D_{10} - D_{00}} \right)^{\beta_1} \right]^{\frac{1}{1 - \beta_1}} > 1.$$

Rearranging gives

$$\beta_1 \left(\frac{D_{10} - D_{11}}{D_{11} - D_{01}} \right) + \left(\frac{D_{11} - D_{00}}{D_{11} - D_{01}} \right)^{\beta_1} < \left(\frac{D_{10} - D_{00}}{D_{11} - D_{01}} \right)^{\beta_1}. \tag{7.100}$$

Substitution of (7.43) and (7.44) in (7.100) gives (7.95).   □

LEMMA 7.5 *It holds that*

$$0 < \arg\min_{Y \geq 0} \gamma(Y) \leq Y_F, \tag{7.101}$$

where the equality sign only holds for $\beta_1 = 1$.

PROOF OF LEMMA 7.5 *From (7.98) we have that $Y^* > 0$. Substitution of equations (7.58) and (7.90) in (7.98) gives*

$$Y^* = Y_F \left[\frac{D_{10} - D_{00}}{(D_{11} - D_{01})^{1-\beta_1}(D_{11} - D_{00})^{\beta_1} + \beta_1(D_{10} - D_{11})} \right]^{\frac{1}{\beta_1 - 1}}.$$

Thus

$$Y^* \leq Y_F,$$

if and only if

$$\left[\frac{D_{10} - D_{00}}{(D_{11} - D_{01})^{1-\beta_1}(D_{11} - D_{00})^{\beta_1} + \beta_1(D_{10} - D_{11})} \right]^{\frac{1}{\beta_1 - 1}} \leq 1.$$

Rewriting gives

$$\frac{D_{10} - D_{00}}{D_{11} - D_{01}} \leq \beta_1 \left(\frac{D_{10} - D_{11}}{D_{11} - D_{01}} \right) + \left(\frac{D_{11} - D_{00}}{D_{11} - D_{01}} \right)^{\beta_1}. \tag{7.102}$$

Combining (7.43) with (7.102) gives

$$f(1) \leq f(\beta_1). \tag{7.103}$$

Define

$$x = \frac{D_{10} - D_{00}}{D_{11} - D_{01}} > 1,$$
$$y = \frac{D_{11} - D_{00}}{D_{11} - D_{01}} < 1.$$

Using these two definitions we have for $\beta_1 \geq 1$:

$$f(\beta_1) = \beta_1(x - y) + y^{\beta_1},$$
$$f'(\beta_1) = x - y + y^{\beta_1}\log(y),$$
$$f''(\beta_1) = y^{\beta_1}(\log(y))^2 > 0.$$

It turns out that f is strictly increasing, because for $0 < y < 1$ we have

$$x > 1 > y - y\log(y) \geq y - y^{\beta_1}\log(y).$$

Thus equation (7.103) holds, the equality sign only holds for $\beta_1 = 1$, and thereby the lemma. □

PROOF OF PROPOSITION 7.4 *Let γ be defined by equation (7.86). If there exists a $Y \in (Y_P, Y_F)$ such that $\gamma(Y) < 0$, the first case applies. If $\gamma(Y) \geq 0$ for all $Y \in (Y_P, Y_F)$ we are in the second case. Lemma 7.2 implies that there exists a $Y \in (Y_P, Y_F)$ such that $\gamma(Y) < 0$ if and only if the minimum of the function γ is negative and reached somewhere between Y_P and Y_F. Lemma 7.4 derives a condition for the minimum of γ to be negative and Lemma 7.5 proves that the minimum is reached in the interval $(0, Y_F)$. Combining Lemmas 7.4 and 7.5 gives Proposition 7.4.* □

LEMMA 7.6 *If*

$$f'(\beta_1) \leq g'(\beta_1), \tag{7.104}$$

then for all $\widehat{\beta_1} \in (\beta_1, \infty)$ it holds that

$$f'\left(\widehat{\beta_1}\right) < g'\left(\widehat{\beta_1}\right). \tag{7.105}$$

PROOF OF LEMMA 7.6 *Define*

$$x = \frac{D_{10} - D_{00}}{D_{11} - D_{01}}, \tag{7.106}$$

$$y = \frac{D_{11} - D_{00}}{D_{11} - D_{01}}. \tag{7.107}$$

Then it holds that $0 < y < 1 < x$ and

$$f'(\beta_1) = x - y + y^{\beta_1} \log(y) > 0, \tag{7.108}$$

$$g'(\beta_1) = x^{\beta_1} \log(x) > 0. \tag{7.109}$$

The proof of $f'(\beta_1)$ being positive is given in the proof of Lemma 7.5. The second and third derivative of f and g are given by

$$f''(\beta_1) = y^{\beta_1} (\log(y))^2 > 0, \tag{7.110}$$

$$f'''(\beta_1) = y^{\beta_1} (\log(y))^3 < 0, \tag{7.111}$$

$$g''(\beta_1) = x^{\beta_1} (\log(x))^2 > 0, \tag{7.112}$$

$$g'''(\beta_1) = x^{\beta_1} (\log(x))^3 > 0. \tag{7.113}$$

First consider the case where $f'(1) > g'(1)$. Due to equations (7.110)-(7.113) we know that f'' is positive and decreasing and g'' is positive and increasing so that there exists a unique β_1^* for which $f'(\beta_1^*) = g'(\beta_1^*)$ and $f'(\beta_1) < g'(\beta_1)$ for all $\beta_1 > \beta_1^*$.

When $f'(1) \leq g'(1)$ we have to prove that for all $\beta_1 > 1$ it holds that $g'(\beta_1) > f'(\beta_1)$. This is certainly true when $f''(\beta_1) < g''(\beta_1)$ for all $\beta_1 \geq 1$. Due to equations (7.111) and (7.113) it is sufficient to prove that $f''(1) < g''(1)$. Thus we have to prove that for $0 < y < 1 < x$,

$$x - y + y \log(y) \leq x \log(x), \tag{7.114}$$

implies

$$y (\log(y))^2 < x (\log(x))^2. \tag{7.115}$$

Using the transformation $u = \log(x)$ and $v = \log(y)$ gives that for $u > 0$ and $v < 0$,

$$e^u (u - 1) - e^v (v - 1) \geq 0, \tag{7.116}$$

has to imply that

$$e^u u^2 - e^v v^2 > 0. \tag{7.117}$$

Consider the (u, v)-plane. Now equation (7.116) holds for a combination of values of u and v on and above the curve $e^u (1 - u) = e^v (1 - v)$ and equation (7.117) holds for u and v values above the curve $e^v v^2 = e^u u^2$. The lemma holds because the curve $e^u (1 - u) = e^v (1 - v)$ is situated above the curve $e^v v^2 = e^u u^2$. This is the case because the curves intersect at $(0,0)$, and the differential $\frac{du}{dv}$ of the first curve is smaller than the corresponding differential of the second curve:

$$\frac{v e^v}{u e^u} < \frac{(v^2 + 2v) e^v}{(u^2 + 2u) e^u}.$$

For a visualization see Figure 7.7.   □

PROOF OF PROPOSITION 7.5 Note that for $\mu > 0$ the relevant β_1 interval is $\left(1, \frac{r}{\mu}\right)$, since

$$\lim_{\sigma \to \infty} \beta_1(\sigma) = 1, \tag{7.118}$$

$$\lim_{\sigma \to 0} \beta_1(\sigma) = \frac{r}{\mu}. \tag{7.119}$$

It holds that

$$f(1) = g(1) = x. \tag{7.120}$$

From (7.108)-(7.110) and (7.112) we know that f and g are convex and increasing in β_1. Further, Lemma 7.6 implies that only the following cases can occur (see also Figure 7.5):

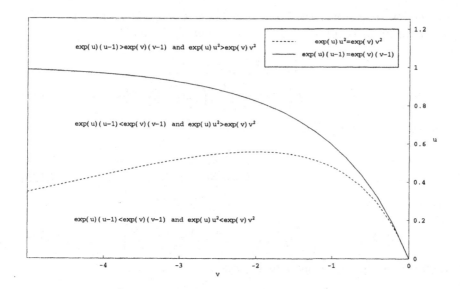

Figure 7.7. The curves $e^u u^2 = e^v v^2$ and $e^u (u - 1) = e^v (v - 1)$ for $u > 0$ and $v < 0$.

1. If $g'(1) \geq f'(1)$ equation (7.45) is satisfied for all $\beta_1 \in \left(1, \frac{r}{\mu}\right)$, so that the equilibrium is always of the preemption type.

2. If $f'(1) > g'(1)$ and $f\left(\frac{r}{\mu}\right) \geq g\left(\frac{r}{\mu}\right)$ equation (7.45) is never satisfied for a $\beta_1 \in \left(1, \frac{r}{\mu}\right)$, thus the equilibrium is always of the joint investment type.

3. If $f'(1) > g'(1)$ and $f\left(\frac{r}{\mu}\right) < g\left(\frac{r}{\mu}\right)$ equation (7.45) is satisfied for high values of β_1 and not satisfied for low values of β_1. $\square$

PROOF OF COROLLARY 7.1 *Recall the quadratic equation (7.8):*

$$Q(\beta_1) = \frac{1}{2}\sigma^2 \beta_1 (\beta_1 - 1) + \mu\beta - r = 0.$$

It holds that

$$\frac{\partial Q}{\partial \beta_1} \frac{\partial \beta_1}{\partial \sigma} + \frac{\partial Q}{\partial \sigma} = 0.$$

Since $\frac{\partial Q}{\partial \beta_1} > 0$ and $\frac{\partial Q}{\partial \sigma} > 0$ we have that $\frac{\partial \beta_1}{\partial \sigma} < 0$. In the same way we can show that $\frac{\partial \beta_1}{\partial \mu} < 0$ and $\frac{\partial \beta_1}{\partial r} > 0$ (see also p. 144 of Dixit and Pindyck, 1996). $\square$

PROOF OF PROPOSITION 7.6 *Since $D_{10} > D_{11}$ we know that $Y_M < Y_J$. Substitution of equation (7.11) in equation (7.75) yields*

$$\phi(Y_M) = I\left[\left(\tfrac{\beta_1}{\beta_1-1}\right)\tfrac{D_{10}-D_{01}}{D_{10}-D_{00}} - 1\right]$$
$$-I\left(\tfrac{D_{11}-D_{01}}{D_{10}-D_{00}}\right)^{\beta_1}\left[\left(\tfrac{\beta_1}{\beta_1-1}\right)\tfrac{D_{10}-D_{01}}{D_{11}-D_{01}} - 1\right]. \quad (7.121)$$

Substitution of the following definitions

$$\xi = \tfrac{D_{10}-D_{01}}{D_{10}-D_{00}} > 1, \quad (7.122)$$
$$\chi = \tfrac{D_{11}-D_{01}}{D_{10}-D_{00}} < 1, \quad (7.123)$$

gives

$$\phi(Y_M) = I\left[\left(\tfrac{\beta_1}{\beta_1-1}\right)\xi\left(1 - \chi^{\beta_1-1}\right) - 1 + \chi^{\beta_1}\right]. \quad (7.124)$$

Differentiating with respect to χ gives

$$\frac{\partial \phi(Y_M)}{\partial \chi} = I\left[-\beta_1 \xi \chi^{\beta_1-2} + \beta_1 \chi^{\beta_1-1}\right]$$
$$= -I\beta_1 \chi^{\beta_1-2}(\xi - \chi) < 0. \quad (7.125)$$

This implies that $\phi(Y_M)$ is decreasing in χ. Since

$$\lim_{\chi \downarrow 1} \phi(Y_M) = 0,$$

we have $\phi(Y_M) \geq 0$. Therefore $Y_P \leq Y_M$. $\square$

PROOF OF PROPOSITION 7.7 *The thresholds Y_L, Y_M, Y_F, and Y_J are decreasing in β_1 since*

$$\frac{\partial \tfrac{\beta_1}{\beta_1-1}}{\partial \beta_1} = -\frac{1}{(\beta_1 - 1)^2} < 0. \quad (7.126)$$

Hence, the only thing that is left to prove is that Y_P decreases with β_1. To do so define for $Y \in [0, Y_F]$ and $\beta_1 \in [0, \infty)$ (cf. (7.65)):

$$\phi(Y, \beta_1) = L(Y) - F(Y)$$
$$= \frac{Y(D_{10}-D_{01})}{r-\mu} - I$$
$$+ \left(\tfrac{Y}{Y_F}\right)^{\beta_1}\left(\tfrac{Y_F(D_{01}-D_{10})}{r-\mu} + I\right). \quad (7.127)$$

From the definition of $Y_P(\beta_1)$ we know that

$$\phi(Y_P(\beta_1), \beta_1) = 0. \quad (7.128)$$

Differentiating (7.128) with respect to β_1 gives

$$\frac{\partial \phi\left(Y, \beta_1\right)}{\partial \beta_1}\bigg|_{Y=Y_P(\beta_1)} + \frac{\partial \phi\left(Y, \beta_1\right)}{\partial Y}\bigg|_{Y=Y_P(\beta_1)} \frac{\partial Y_P\left(\beta_1\right)}{\partial \beta_1} = 0. \qquad (7.129)$$

Hence, to say something about the sign of $\frac{\partial Y_P(\beta_1)}{\partial \beta_1}$, we need to determine the signs of $\frac{\partial \phi(Y,\beta_1)}{\partial \beta_1}\big|_{Y=Y_P(\beta_1)}$ and $\frac{\partial \phi(Y,\beta_1)}{\partial Y}\big|_{Y=Y_P(\beta_1)}$. From Lemma 7.1 we already know that

$$\frac{\partial \phi\left(Y, \beta_1\right)}{\partial Y}\bigg|_{Y=Y_P(\beta_1)} > 0. \qquad (7.130)$$

Now let us concentrate at $\frac{\partial \phi(Y,\beta_1)}{\partial \beta_1}\big|_{Y=Y_P(\beta_1)}$. To do so, first substitute equation (7.18) in (7.127), which gives

$$\phi\left(Y, \beta_1\right) = \frac{Y\left(D_{10} - D_{01}\right)}{r - \mu} - I \qquad (7.131)$$
$$+ \frac{I\left(Y \frac{\beta_1 - 1}{\beta_1} \frac{D_{11} - D_{01}}{(r - \mu)I}\right)^{\beta_1}}{(\beta_1 - 1)(D_{11} - D_{01})}\left((\beta_1 - 1) D_{11} - \beta_1 D_{10} + D_{01}\right).$$

From (7.131) it is obtained that

$$\frac{\partial \phi\left(Y, \beta_1\right)}{\partial \beta_1} = \frac{I\left(Y \frac{\beta_1 - 1}{\beta_1} \frac{D_{11} - D_{01}}{(r - \mu)I}\right)^{\beta_1}}{(\beta_1 - 1)(D_{11} - D_{01})}\left(D_{11} - D_{10}\right) \qquad (7.132)$$
$$+ \log\left(Y \frac{\beta_1 - 1}{\beta_1} \frac{D_{11} - D_{01}}{(r - \mu)I}\right)\left((\beta_1 - 1) D_{11} - \beta_1 D_{10} + D_{01}\right).$$

Define

$$\overline{Y}\left(\beta_1\right) = \frac{\beta_1}{\beta_1 - 1} \frac{(r - \mu) I}{D_{10} - D_{01}}. \qquad (7.133)$$

Substitution of (7.133) into (7.131) gives

$$\phi\left(\overline{Y}\left(\beta_1\right), \beta_1\right) = \frac{I}{\beta_1 - 1}\left[1 + \left(\frac{D_{11} - D_{01}}{D_{10} - D_{01}}\right)^{\beta_1}\left(\beta_1 - 1 - \beta_1\left(\frac{D_{10} - D_{01}}{D_{11} - D_{01}}\right)\right)\right]$$
$$= \frac{I}{\beta_1 - 1}\left[1 + x^{\beta_1}\left(\beta_1 - 1 - \frac{\beta_1}{x}\right)\right] > 0, \qquad (7.134)$$

where $x = \frac{D_{11} - D_{01}}{D_{10} - D_{01}}$ and $x \in (0,1)$. Lemma 7.1 and equation (7.134) imply that

$$Y_P\left(\beta_1\right) < \overline{Y}\left(\beta_1\right). \qquad (7.135)$$

From (7.132) we conclude that $\frac{\partial \phi(Y,\beta_1)}{\partial \beta_1} > 0$ for sufficiently low values of Y.

It holds that

$$\frac{\partial \phi(Y,\beta_1)}{\partial \beta_1}\bigg|_{Y=\overline{Y}(\beta_1)} > 0, \qquad (7.136)$$

if and only if

$$\varphi(\beta_1) = D_{11} - D_{10}$$
$$+ \log\left(\frac{D_{11}-D_{01}}{D_{10}-D_{01}}\right)((\beta_1 - 1)D_{11} - \beta_1 D_{10} + D_{01})$$
$$> 0. \qquad (7.137)$$

To prove this we first note that the function $\varphi(\beta_1)$ is increasing in β_1 :

$$\frac{\partial \varphi(\beta_1)}{\partial \beta_1} = \log\left(\frac{D_{11}-D_{01}}{D_{10}-D_{01}}\right)(D_{11} - D_{10}) > 0.$$

Furthermore $\varphi(1) > 0$, since

$$\log\left(\frac{D_{11}-D_{01}}{D_{10}-D_{01}}\right) < \frac{D_{11} - D_{01}}{D_{10} - D_{01}} - 1,$$

so that (7.137) is valid. Thus equation (7.136) holds. Now, from (7.135) and (7.136) we have

$$\frac{\partial \phi(Y,\beta_1)}{\partial \beta_1}\bigg|_{Y=Y_P(\beta_1)} > 0. \qquad (7.138)$$

Finally, from (7.129), (7.130), and (7.138) it can be concluded that

$$\frac{\partial Y_P(\beta_1)}{\partial \beta_1} < 0.$$

$\square$

PROOF OF PROPOSITION 7.8 *The option value of waiting in the monopoly case is given by equation (7.14). At the moment of investment in the preemption case, the strategic option value of waiting equals*

$$L(Y_P) - \frac{Y_P D_{00}}{r-\mu} < \frac{Y_P D_{10}}{r-\mu} - I - \frac{Y_P D_{00}}{r-\mu}$$
$$< \frac{Y_M(D_{10} - D_{00})}{r-\mu} - I$$
$$= \frac{I}{\beta_1 - 1}.$$

In the joint investment case we have

$$J\left(Y_J, Y_J\right) - \frac{Y_J D_{00}}{r - \mu} = \frac{Y_J D_{11}}{r - \mu} - I - \frac{Y_J D_{00}}{r - \mu}$$

$$= \frac{I}{\beta_1 - 1}.$$

□

Chapter 8

ONE NEW TECHNOLOGY
AND ASYMMETRIC FIRMS

1.   INTRODUCTION

In Nielsen, 1999 and in Chapter 7 it is shown that, in a strategic investment new market model, competition by an identical firm precipitates investment. The purpose of this chapter is to examine the same issue, namely the effect of introducing another firm on the original firm's investment decision, in an asymmetric setting. We introduce asymmetry by letting the firms have different investment costs, but the methods and results should be extendable to other types of asymmetry as well.

We find that competition precipitates investment in an asymmetric setting as well, but in a weaker sense. More precisely, if the investment cost of the new firm is sufficiently low, competition strictly precipitates investment, but if the investment cost is high, the introduction of the new firm does not have an effect on the investment strategy of the old firm. This result holds both when there are negative or positive externalities to investment. Though, the type of externality influences the critical investment cost level.

In Chapters I.4 and II.3 of Torvund, 1999 almost the same model is considered. He only analyzes the negative externalities case and makes explicit assumptions on the demand and supply functions of the market. Torvund does not explicitly state the strategies that result in the equilibria. His conclusions coincide with ours for the negative externalities case.

The remainder of the chapter is organized as follows. In Section 2 we present the model. The value functions and investment thresholds are derived in Section 3. The negative externalities case is analyzed in

Section 4. Section 5 deals with the positive externalities case and Section 6 concludes.

2.   THE MODEL

We consider two risk-neutral firms that can make an irreversible investment in order to become active on a new market. The firms maximize their value over an infinite planning horizon and discount at rate $r\,(>0)$. We denote the firms by i and j, with $i,j \in \{1,2\}$ and $i \neq j$. The profit flow of firm i at time $t\,(\geq 0)$ equals

$$Y\,(t)\,D_{N_i N_j}, \tag{8.1}$$

where, for $k \in \{i,j\}$:

$$N_k = \begin{cases} 0 & \text{if firm } k \text{ has not invested,} \\ 1 & \text{if firm } k \text{ has invested.} \end{cases} \tag{8.2}$$

$Y\,(t)$ behaves according to the following geometric Brownian motion process:

$$dY\,(t) = \mu Y\,(t)\,dt + \sigma Y\,(t)\,d\omega\,(t), \tag{8.3}$$
$$Y\,(0) = y, \tag{8.4}$$

where $y > 0$, $0 < \mu < r$, $\sigma > 0$, and the $d\omega\,(t)$'s are independently and identically distributed according to a normal distribution with mean zero and variance dt. Since we consider a new market, we set $D_{00} = D_{01} = 0$. The investment cost for firm i equals I_i, with $i \in \{1,2\}$. We assume without loss of generality that $I_2 > I_1 > 0$.

3.   VALUE FUNCTIONS AND INVESTMENT THRESHOLDS

We solve the model described in Section 2 using the game theoretic concept of timing games. The approach applied has been introduced in Appendix A of Chapter 4. In a timing game the players must decide when to make a single move. The player that moves first is called leader, and the other is the follower. Players can also decide to move simultaneously. First we introduce some more notation. We denote by $L_j\,(Y\,(t))$ the payoff at time t to player j if none of the players has moved before time t and player j moves alone at time t. The payoff of the follower then is denoted by $F_i\,(Y\,(t))$. When both players move simultaneously at time t their payoffs equal $M_1\,(Y\,(t))$ for firm 1 and $M_2\,(Y\,(t))$ for firm 2. In the remainder of this chapter we omit the time dependence of Y when there is no confusion possible.

Dynamic games are usually solved backwards and this one is no exception. We start with deriving the leader, follower and joint investment curves for the model. Subsection 3.1 is devoted to the value of being follower, Subsection 3.2 to the value of being leader, and the value of joint investment is treated in Subsection 3.3.

3.1 FOLLOWER

Without loss of generality assume that firm j has and firm i has not invested. The problem facing the follower is an optimal stopping problem (see Appendix A of Chapter 2), which conjectures the existence of a threshold Y_{F_i} such that investing is optimal for firm i whenever $Y \geq Y_{F_i}$ and waiting is optimal otherwise. Solving this optimal stopping problem, which is a simplification of the one presented and solved in Appendix A of Chapter 7, gives rise to the following value function for firm i

$$F_i(Y) = \begin{cases} A_{i1} Y^{\beta_1} & \text{if } Y < Y_{F_i}, \\ \frac{Y D_{11}}{r-\mu} - I_i & \text{if } Y \geq Y_{F_i}, \end{cases} \tag{8.5}$$

where β_1 is the positive root of the following quadratic equation

$$\frac{1}{2}\sigma^2 \beta^2 + \left(\mu - \frac{1}{2}\sigma^2\right)\beta - r = 0. \tag{8.6}$$

Solving the value matching and smooth pasting conditions (see Appendix A of Chapter 2) simultaneously gives the following expressions for the threshold and the constant

$$Y_{F_i} = \frac{\beta_1}{\beta_1 - 1} \frac{(r-\mu) I_i}{D_{11}}, \tag{8.7}$$

$$A_{i1} = \frac{Y_{F_i}^{1-\beta_1}}{\beta_1} \frac{D_{11}}{r-\mu}. \tag{8.8}$$

Using the same steps as in the proof of Proposition 7.1 one can prove that Y_{F_i} is unique. The optimal investment time T_{F_i} of firm i as follower is equal to

$$T_{F_i} = \inf\left(t \mid Y(t) \geq Y_{F_i}\right). \tag{8.9}$$

3.2 LEADER

Firm j, being the leader, knows the optimal response of firm i on its investment at time t. The value of firm j at $Y(t) = Y$ if it invests at time t equals

$$L_j(Y) = E\left[\int_{\tau=t}^{\max(T_{F_i},t)} Y(\tau)D_{10}e^{-r(\tau-t)}d\tau - I_j\right.$$

$$\left. + \int_{\tau=\max(T_{F_i},t)}^{\infty} Y(\tau)D_{11}e^{-r(\tau-t)}d\tau \,\middle|\, Y(t) = Y\right].(8.10)$$

Rewriting gives (see Appendix A.2 of Chapter 7 for details)

$$L_j(Y) = \begin{cases} \frac{YD_{10}}{r-\mu} - I_j + \left(\frac{Y}{Y_{F_i}}\right)^{\beta_1}\frac{Y_{F_i}(D_{11}-D_{10})}{r-\mu} & \text{if } Y < Y_{F_i}, \\ \frac{YD_{11}}{r-\mu} - I_j & \text{if } Y \geq Y_{F_i}. \end{cases} \quad (8.11)$$

3.3   JOINT INVESTMENT

The value of firm $i \in \{1,2\}$ if both firms invest simultaneously is given by

$$M_i(Y) = \frac{YD_{11}}{r-\mu} - I_i. \qquad (8.12)$$

The optimal joint investment time for firm i equals T_{F_i}, which thus equals the follower's threshold. The new market assumption is the reason for this. Assume for a moment that both firms are already active on the output market before the first investment is made. Then the investment of the first firm decreases the profit flow of the second firm. This decrease in its profit flow gives the second firm an incentive to make its investment earlier. The reason is that the gain of the investment is larger. For a more formal treatment and proof of this phenomenon see Chapter 7.

4.   NEGATIVE EXTERNALITIES

In the negative externalities case the firms compete in the traditional sense. While we do not model the product market explicitly, the assumption that profit flow falls upon investment by a second firm is compatible with both a fall in market share and a fall in price due to increase in supply. In the model the negative externalities case is characterized by the following equation

$$D_{10} > D_{11}. \qquad (8.13)$$

As in the previous chapter the equilibria of the investment game depend on the relative positions of the leader, follower and joint investment curves of each firm. The following proposition gives the three different cases for the curves of firm 2 that can occur. The proof is given in Appendix A.

PROPOSITION 8.1 *Let*

$$I_2^* = \frac{I_1}{D_{11}} \left(\frac{D_{10}^{\beta_1} - D_{11}^{\beta_1}}{\beta_1 (D_{10} - D_{11})} \right)^{\frac{1}{\beta_1 - 1}}. \tag{8.14}$$

Then it holds that $I_2^* > I_1$ and moreover,

1. *if $I_2 \in (I_2^*, \infty)$ it holds that $L_2(Y) < F_2(Y)$ for all $Y \in (0, Y_{F_2})$ (case 1, see Figure 8.1);*

2. *if $I_2 = I_2^*$ it holds that $L_2(Y_{P_2}) = F_2(Y_{P_2})$ for some unique $Y_{P_2} \in (0, Y_{F_2})$ (case 2, see Figure 8.2);*

3. *if $I_2 \in (I_1, I_2^*)$ it holds that $L_2(Y) > F_2(Y)$ for all $Y \in (Y_{P_{21}}, Y_{P_{22}})$ and $L_2(Y) < F_2(Y)$ for all $Y \in (0, Y_{P_{21}}) \cup (Y_{P_{22}}, Y_{F_2})$ with $0 < Y_{P_{21}} < Y_{P_{22}} < Y_{F_1}$ (case 3, see Figure 8.3).*

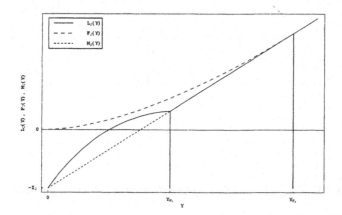

Figure 8.1. Leader, follower and joint investment curves of firm 2 in case 1.

In cases 1 and 2 the investment cost of firm 2 is, relatively to the investment cost of firm 1, that high that firm 2's leader curve never exceeds its follower curve. For Y values larger than or equal to Y_{F_1} the leader value of the second firm equals its joint investment curve. From Y_{F_2} on the three payoff functions coincide. Consequently, firm 2 never becomes leader in any of these two cases and always invests at time T_{F_2}.

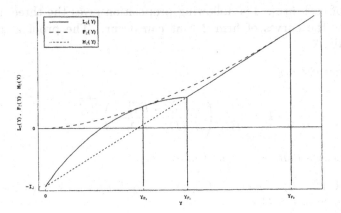

Figure 8.2. Leader, follower and joint investment curves of firm 2 in case 2.

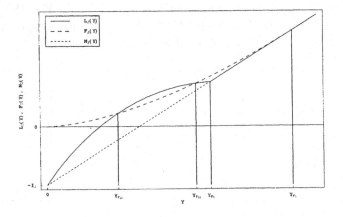

Figure 8.3. Leader, follower and joint investment curves of firm 2 in case 3.

In case 3 the investment costs of both firms are almost similar. It turns out that there exists a Y for which the firm 2's leader curve exceeds its follower curve. For $Y = 0$ as well as for $Y = Y_{F_2}$ the follower value of firm 2 exceeds the leader value. These observations imply the existence of the preemption interval $(Y_{P_{21}}, Y_{P_{22}})$, which on its turn results in the following corollary.

COROLLARY 8.1 *Firm 2 has only incentives to preempt in case 3.*

It turns out that firm 1 has always an incentive to preempt. This is formally stated in Proposition 8.2, which is proved in Appendix A.

PROPOSITION 8.2 *It holds that* $L_1(Y) > F_1(Y)$ *for* $Y \in (Y_{P_1}, Y_{F_2})$ *with* $Y_{P_1} \in (0, Y_{F_1})$ *(see Figure 8.4).*

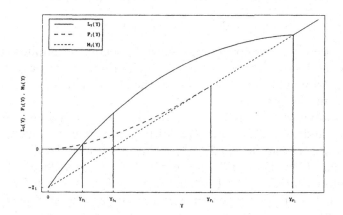

Figure 8.4.   Leader, follower and joint investment curves of firm 1.

As stated above, in cases 1 and 2 of Proposition 8.1 firm 2 will never try to preempt firm 1 with its investment. Therefore firm 1 can optimize its investment time without needing to take into account the investment strategy of firm 2. Therefore this equilibrium is similar to the open loop equilibrium of the game (see Chapter 4). Solving the optimal stopping problem that firm 1 faces (see Appendix A.2 of Chapter 7) leads to the following investment threshold

$$Y_{L_1} = \frac{\beta_1}{\beta_1 - 1} \frac{(r - \mu) I_1}{D_{10}}. \qquad (8.15)$$

This threshold leads to the following optimal investment time for firm 1

$$T_{L_1} = \inf(t \,|\, Y(t) \geq Y_{L_1}). \qquad (8.16)$$

Note that equations (8.7), (8.13), and (8.15) imply that $Y_{L_1} < Y_{F_1} < Y_{F_2}$, so that (indeed) it holds that $T_{L_1} \leq T_{F_2}$.

Firm 2 will try to preempt firm 1 in case 3, whenever firm 1 has not invested before time

$$T_{P_{21}} = \inf(t \,|\, Y(t) \in [Y_{P_{21}}, Y_{P_{22}}]), \qquad (8.17)$$

since for firm 2 the leader curve exceeds its follower curve on the interval $(Y_{P_{21}}, Y_{P_{22}})$.

PROPOSITION 8.3 *There exists a unique* $I_2^{**} \in (I_1, I_2^*)$ *such that*

1. $Y_{L_1} > Y_{P_{21}}$ *if and only if* $I_2 \in (I_1, I_2^{**})$ *(case 3.1);*

2. $Y_{L_1} \leq Y_{P_{21}}$ *if and only if* $I_2 \in [I_2^{**}, I_2^*)$ *(case 3.2).*

Proposition 8.3 (the proof is given in Appendix A) implies that the equilibrium of case 3.2 is equal to the equilibrium of the cases 1 and 2. The equilibrium of case 3.1 depends on the initial value of the geometric Brownian motion process.

If the initial value y is below or equal to $Y_{P_{21}}$ firm 1 will invest at time $T_{P_{21}}$ and firm 2 at time T_{F_2}. Firm 1 is willing to invest first at any level larger than or equal to Y_{P_1}, since its leader payoff exceeds its follower payoff for these Y's. Since firm 1 knows that firm 2 does not invest before time $T_{P_{21}}$ firm 1 will not invest before that time $T_{P_{21}}$ either. For firm 1 it is optimal to invest with probability one at time $T_{P_{21}}$ and consequently firm 2 does not invest at that time but at time T_{F_2}.

For an initial value $y \in (Y_{P_{21}}, Y_{P_{22}})$ both firms want to become leader and therefore both firms invest with positive probability at time $t = 0$. The result is that the probability that both firms invest simultaneously is strictly positive. The exact probability that a firm becomes leader or follower and the probability that there is joint investment are given below.

Firm 2's follower payoff exceeds its leader payoff for $y \in [Y_{P_{22}}, Y_{F_2})$ and therefore firm 2 invests at time T_{F_2} and since $Y_{L_1} \leq Y_{P_{22}}$ (see equations (8.15) and (8.36)) firm 1 invests at time $t = 0$.

It is optimal for both firms to invest at time $t = 0$ if the initial value y is larger than or equal to Y_{F_2}.

All the equilibria are summarized in the following theorem. The equilibrium strategies that lead to the equilibria are given in Appendix B.1.

THEOREM 8.1 *For* $y \geq 0$, *let*

$$\alpha_1(y) = \frac{L_2(y) - F_2(y)}{L_2(y) - M_2(y)}, \tag{8.18}$$

$$\alpha_2(y) = \frac{L_1(y) - F_1(y)}{L_1(y) - M_1(y)}. \tag{8.19}$$

The equilibrium outcome in the negative externalities case is as follows:

1. *if* $I_2 \in [I_2^{**}, \infty)$ *(with probability one) firm 1 invests at time* T_{L_1} *and firm 2 invests at time* T_{F_2};

2. *if* $I_2 \in (I_1, I_2^{**})$ *the equilibrium outcome depends on the initial value of the geometric Brownian motion process:*

(a) if $y \leq Y_{P_{21}}$ *(with probability one) firm 1 invests at time* $T_{P_{21}}$ *and firm 2 invests at time* T_{F_2};

(b) if $Y_{P_{21}} < y < Y_{P_{22}}$ *there are three possible outcomes:*

   i. *with probability* $\frac{\alpha_1(y)(1-\alpha_2(y))}{\alpha_1(y)+\alpha_2(y)-\alpha_1(y)\alpha_2(y)}$, *firm 1 invests at time 0 and firm 2 invests at time* T_{F_2};

   ii. *with probability* $\frac{\alpha_2(y)(1-\alpha_1(y))}{\alpha_1(y)+\alpha_2(y)-\alpha_1(y)\alpha_2(y)}$, *firm 1 invests at time* T_{F_1} *and firm 2 invests at time 0;*

   iii. *with probability* $\frac{\alpha_1(y)\alpha_2(y)}{\alpha_1(y)+\alpha_2(y)-\alpha_1(y)\alpha_2(y)}$, *both firm 1 and firm 2 invest at time 0;*

(c) if $Y_{P_{22}} \leq y < Y_{F_2}$ *(with probability one) firm 1 invests at time 0 and firm 2 invests at time* T_{F_2};

(d) if $y \geq Y_{F_2}$ *(with probability one) both firm 1 and firm 2 invest at time 0.*

5.   POSITIVE EXTERNALITIES

In some situations an investment is more profitable when more firms have invested. This situation could arise if the firms produce complementary products or if there are network externalities. The positive externalities case is characterized by

$$D_{11} \geq D_{10}. \tag{8.20}$$

Compared to the negative externalities case, the thresholds Y_{L_i} and Y_{F_i} switch places ($i \in \{1,2\}$). The reason is that joint investment is more attractive than single investment in this section. To derive the equilibria we need to know whether the leader threshold of firm i is smaller than the follower threshold of firm j. The following proposition states the possible cases. The proofs are given in Appendix A.

PROPOSITION 8.4 *Let*

$$I_1^* = I_2 \frac{D_{10}}{D_{11}}. \tag{8.21}$$

Then it holds that $I_1^* \leq I_2$ *and moreover,*

1. *if* $I_1 \in (0, I_1^*)$ *it holds that* $Y_{L_1} < Y_{F_2}$ *(case 4, see Figure 8.5), where* Y_{L_1} *and* Y_{F_2} *are given by equations (8.15) and (8.7), respectively;*

2. *if* $I_1 \in [I_1^*, I_2)$ *it holds that* $Y_{L_1} \geq Y_{F_2}$ *(case 5, see Figure 8.6).*

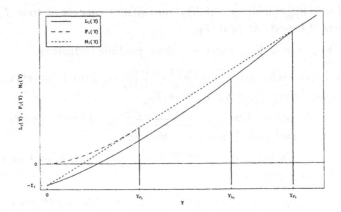

Figure 8.5. Leader, follower and joint investment curves of firm 1 in case 4.

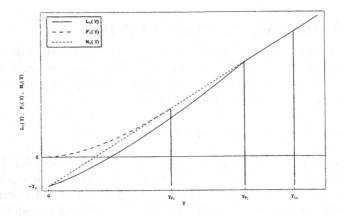

Figure 8.6. Leader, follower and joint investment curves of firm 1 in case 5.

Due to equation (8.20) firm i's leader curve is situated below its follower curve for all $Y \in [0, Y_{F_j})$. In case 4 the investment cost of firm 1 is low enough to trigger investment by firm 1 before time T_{F_2}. This contrary to case 5 where firm 1's leader trigger is larger than firm 2's follower trigger.

The following proposition states that the follower threshold of firm 1 is always smaller than firm 2's leader threshold.

PROPOSITION 8.5 *It holds that* $Y_{L_2} > Y_{F_1}$ *(see Figure 8.7), where* Y_{F_1} *is defined by equation (8.7) and* Y_{L_2} *equals*

$$Y_{L_2} = \frac{\beta_1}{\beta_1 - 1} \frac{(r - \mu) I_2}{D_{10}}. \tag{8.22}$$

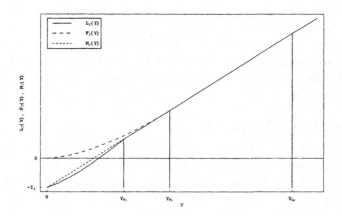

Figure 8.7. Leader, follower and joint investment curves of firm 2.

From the propositions it follows that in case 4 firm 1 invests at time T_{L_1}, although firm 1 prefers to become follower there: $L_1(Y_{L_1}) < F_1(Y_{F_1})$. The reason is that firm 1 knows for sure that firm 2 is not going to invest before time T_{F_2}. Therefore firm 1 can choose between delaying its investment and investing at T_{L_1}. Due to the definition of Y_{L_1} it turns out that investing at time T_{L_1} is the optimal action for firm 1. Notice that the game is an attrition game for which there does not exist a symmetric equilibrium (see Appendix A.3 of Chapter 4). Firm 2 invests at time T_{F_2}.

In case 5 the firms invest simultaneously at time T_{F_2}, one firm initiates the investment and the other will make it joint investment. The following theorem formally states the equilibria outcomes (in Appendix B.2 the equilibrium strategies are presented).

THEOREM 8.2 *The equilibrium outcome in the positive externalities case is as follows:*

1. *if* $I_1 \in (0, I_1^*)$ *(with probability one) firm 1 invests at time* T_{L_1} *and firm 2 invests at time* T_{F_2};

2. *if $I_1 \in [I_1^*, I_2)$ (with probability one) both firm 1 and firm 2 invest at time T_{F_2}.*

Note that it is not clear which of the firms has the highest payoff in case 4. There are two opposite effects. On the one hand firm 1 invests first and therefore gets a payoff equal to its leader value, which is lower than its follower payoff, and firm 2 gets its follower payoff, which is higher than its leader payoff. On the other hand the investment costs of firm 1 are lower than those of firm 2 which implies that all the payoff curves of firm 1 are situated above those of firm 2. It is not clear which effect dominates.

6.   CONCLUSIONS

We are now in a position to state our main result. Let us say, that competition precipitates investment if the first investment in a two firm model is never made later than in the corresponding one firm model. Then we have the following theorem.

THEOREM 8.3 *For any D_{10}, D_{11} and any I_1, I_2 competition precipitates investment.*

At this point, it may be useful to briefly review the arguments leading to the conclusion that competition weakly precipitates investment even when firms are allowed to be asymmetric.

We found that, when firm 2's investment costs are very high, its presence has no strategic effect on the investment behavior of firm 1. Firm 1 simply proceeds and invests at its most preferred investment threshold. Crucially, this most preferred investment threshold turns out to be the same as in the model in which firm 1 is the only firm in both the case of negative and positive externalities.

We proceeded to analyze the cases in which firm 2 has low enough investment costs for its presence to have a strategic effect. In both the case of negative and positive externalities, this effect turns out to precipitate investment, but for very different reasons. When there are negative externalities, the threat of preemption pushes firm 1 to invest earlier than it would otherwise have done. When there are positive externalities, both firms invest early in anticipation that the other firm will invest early as well.

Appendices

A.   LEMMA AND PROOFS

LEMMA 8.1 *For $u > v > 0$ and $a > 1$ it holds that*

$$u^a - v^a - auv^{a-1} + av^a > 0. \tag{8.23}$$

PROOF OF LEMMA 8.1 *Equation (8.23) holds if and only if (divide by v^a)*

$$\left(\frac{u}{v}\right)^a - 1 - a\left(\frac{u}{v}\right) + a > 0.$$

Define $w = \frac{u}{v}$ and for $w \geq 1$ the function $g(w) = w^a - 1 - aw + a$. It is easy to check that for $w > 1$

$$\frac{\partial g(w)}{\partial w} = aw^{a-1} - a > 0,$$

and $g(1) = 0$. Thus equation (8.23) and thereby the lemma holds.   □

PROOF OF PROPOSITION 8.1 *Define the function $\phi_2 : [0, Y_{F_1}] \to \mathbf{R}$ as follows*

$$\phi_2(Y) = L_2(Y) - F_2(Y). \tag{8.24}$$

Substitution of equations (8.5) and (8.11) into the last equation gives

$$
\begin{aligned}
\phi_2(Y) &= \frac{Y D_{10}}{r - \mu} - I_2 + \left(\frac{Y}{Y_{F_1}}\right)^{\beta_1} \frac{Y_{F_1}(D_{11} - D_{10})}{r - \mu} \\
&\quad - \left(\frac{Y}{Y_{F_2}}\right)^{\beta_1} \frac{Y_{F_2} D_{11}}{\beta_1(r - \mu)}.
\end{aligned} \tag{8.25}
$$

Then it follows that

$$
\begin{aligned}
\phi_2(0) &= -I_2 < 0, \tag{8.26} \\
\phi_2(Y_{F_1}) &= \frac{Y_{F_1} D_{11}}{r - \mu} - I_2 - \left(\frac{Y_{F_1}}{Y_{F_2}}\right)^{\beta_1} \frac{Y_{F_2} D_{11}}{\beta_1(r - \mu)} < 0, \tag{8.27}
\end{aligned}
$$

where the last inequality sign is a direct result of the definition of Y_{F_2}. The first and second derivative of ϕ_2 are given by

$$\frac{\partial \phi_2 (Y)}{\partial Y} = \frac{D_{10}}{r - \mu} + \beta_1 Y^{\beta_1 - 1} \frac{Y_{F_1}^{1-\beta_1} (D_{11} - D_{10})}{r - \mu}$$
$$- \beta_1 Y^{\beta_1 - 1} \frac{Y_{F_2}^{1-\beta_1} D_{11}}{\beta_1 (r - \mu)}, \tag{8.28}$$

$$\frac{\partial^2 \phi_2 (Y)}{\partial Y^2} = \beta_1 (\beta_1 - 1) Y^{\beta_1 - 2} \frac{Y_{F_1}^{1-\beta_1} (D_{11} - D_{10})}{r - \mu}$$
$$- \beta_1 (\beta_1 - 1) Y^{\beta_1 - 2} \frac{Y_{F_2}^{1-\beta_1} D_{11}}{\beta_1 (r - \mu)} < 0. \tag{8.29}$$

Thus ϕ_2 is strictly concave. Solving for the maximum yields

$$Y_{P_2} = \frac{\beta_1}{\beta_1 - 1} \frac{(r - \mu) I_2}{D_{10}}. \tag{8.30}$$

From equations (8.7), (8.13), and (8.30) it follows that $Y_{P_2} \in (0, Y_{F_1})$. Substituting the expressions for Y_{P_2}, Y_{F_1}, and Y_{F_2} in $\phi_2 (Y_{P_2})$ and set this value equal to zero gives

$$\frac{\beta_1 I_2}{\beta_1 - 1} - I_2 + \left(\frac{D_{11} I_2}{D_{10} I_1} \right)^{\beta_1} \frac{\beta_1 I_1 (D_{11} - D_{10})}{(\beta_1 - 1) D_{11}} - \left(\frac{D_{11}}{D_{10}} \right)^{\beta_1} \frac{I_2}{\beta_1 - 1} = 0. \tag{8.31}$$

Solving for I_2 yields the expression for I_2^* :

$$I_2^* = \frac{I_1}{D_{11}} \left(\frac{D_{10}^{\beta_1} - D_{11}^{\beta_1}}{\beta_1 (D_{10} - D_{11})} \right)^{\frac{1}{\beta_1 - 1}}. \tag{8.32}$$

Therefore the three cases of Proposition 8.1 apply. The last step is to prove that $I_2^* > I_1$. From equation (8.32) it follows that $I_2^* > I_1$ if and only if

$$\left(\frac{D_{10}^{\beta_1} - D_{11}^{\beta_1}}{\beta_1 (D_{10} - D_{11})} \right)^{\frac{1}{\beta_1 - 1}} > D_{11}. \tag{8.33}$$

Rewriting equation (8.33) leads to

$$D_{10}^{\beta_1} - D_{11}^{\beta_1} - \beta_1 D_{10} D_{11}^{\beta_1 - 1} + \beta_1 D_{11}^{\beta_1} > 0. \tag{8.34}$$

Due to Lemma 8.1 with $u = D_{10}$, $v = D_{11}$, and $a = \beta_1$, equation (8.34) always holds. $\qquad\square$

PROOF OF PROPOSITION 8.2 *Define the function* $\phi_1 : [0, Y_{F_2}] \to \mathbf{R}$ *as follows*

$$\phi_1(Y) = L_1(Y) - F_1(Y). \tag{8.35}$$

Substitution of equations (8.5) and (8.11) into equation (8.35) gives

$$\phi_1(Y) = \begin{cases} \frac{Y D_{10}}{r-\mu} - I_1 + \left(\frac{Y}{Y_{F_2}}\right)^{\beta_1} \frac{Y_{F_2}(D_{11}-D_{10})}{r-\mu} \\ \quad - \left(\frac{Y}{Y_{F_1}}\right)^{\beta_1} \frac{Y_{F_1} D_{11}}{\beta_1(r-\mu)} & \text{if } Y \in [0, Y_{F_1}), \\ \frac{Y(D_{10}-D_{11})}{r-\mu} + \left(\frac{Y}{Y_{F_2}}\right)^{\beta_1} \frac{Y_{F_2}(D_{11}-D_{10})}{r-\mu} & \text{if } Y \in [Y_{F_1}, Y_{F_2}]. \end{cases} \tag{8.36}$$

The first and second derivative of ϕ_1 are equal to

$$\frac{\partial \phi_1(Y)}{\partial Y} = \begin{cases} \frac{D_{10}}{r-\mu} + \beta_1 Y^{\beta_1-1} \frac{Y_{F_2}^{1-\beta_1}(D_{11}-D_{10})}{r-\mu} \\ \quad -\frac{D_{10}}{r-\mu} + \beta_1 Y^{\beta_1-1} \frac{Y_{F_1}^{1-\beta_1} D_{11}}{\beta_1(r-\mu)} & \text{if } Y \in [0, Y_{F_1}), \\ \frac{D_{10}-D_{11}}{r-\mu} + \beta_1 Y^{\beta_1-1} \frac{Y_{F_2}^{1-\beta_1}(D_{11}-D_{10})}{r-\mu} & \text{if } Y \in [Y_{F_1}, Y_{F_2}], \end{cases} \tag{8.37}$$

$$\frac{\partial^2 \phi_1(Y)}{\partial Y^2} = \begin{cases} \beta_1(\beta_1-1) Y^{\beta_1-2} \frac{Y_{F_2}^{1-\beta_1}(D_{11}-D_{10})}{r-\mu} \\ \quad -\beta_1(\beta_1-1) Y^{\beta_1-2} \frac{Y_{F_1}^{1-\beta_1} D_{11}}{\beta_1(r-\mu)} & \text{if } Y \in (0, Y_{F_1}), \\ \beta_1(\beta_1-1) Y^{\beta_1-2} \frac{Y_{F_2}^{1-\beta_1}(D_{11}-D_{10})}{r-\mu} & \text{if } Y \in (Y_{F_1}, Y_{F_2}]. \end{cases} \tag{8.38}$$

The last equation implies that ϕ_1 is strictly concave on the interval $[0, Y_{F_2}]$. Due to the derivation of Y_{F_1} we know that

$$\lim_{Y \uparrow Y_{F_1}} \phi_1(Y) = \lim_{Y \downarrow Y_{F_1}} \phi_1(Y), \tag{8.39}$$

$$\lim_{Y \uparrow Y_{F_1}} \frac{\partial \phi_1(Y)}{\partial Y} = \lim_{Y \downarrow Y_{F_1}} \frac{\partial \phi_1(Y)}{\partial Y}. \tag{8.40}$$

Further, it holds that

$$\phi_1(0) = -I_1 < 0, \tag{8.41}$$

$$\phi_1(Y_{F_2}) = 0, \tag{8.42}$$

$$\left. \frac{\partial \phi_1(Y)}{\partial Y} \right|_{Y=Y_{F_2}} = (\beta_1 - 1) \frac{D_{11} - D_{10}}{r-\mu} < 0. \tag{8.43}$$

Equations (8.41), (8.42), and (8.43) imply the existence of Y_{P_1}. The uniqueness follows from the strict concavity of ϕ_1.   ☐

PROOF OF PROPOSITION 8.3 *It holds that $Y_{L_1} \leq Y_{P_{21}}$ if and only if $F_2(Y_{L_1}) \geq L_2(Y_{L_1})$. Let ϕ_2 be given by equation (8.24). Then we have that*

$$\phi_2(Y_{L_1}) = \frac{\beta_1 I_1}{\beta_1 - 1} - I_2 + \left(\frac{D_{11}}{D_{10}}\right)^{\beta_1} \frac{\beta_1 I_1 (D_{11} - D_{10})}{(\beta_1 - 1) D_{11}} - \left(\frac{I_1 D_{11}}{I_2 D_{10}}\right)^{\beta_1} \frac{I_2}{\beta_1 - 1}.$$
(8.44)

Multiplying by $\frac{\beta_1 - 1}{I_2} D_{10}^{\beta_1}$ gives

$$\frac{I_1}{I_2} \beta_1 D_{10}^{\beta_1} - (\beta_1 - 1) D_{10}^{\beta_1} + D_{11}^{\beta_1 - 1} \frac{I_1}{I_2} \beta_1 (D_{10} - D_{11}) - \left(\frac{I_1}{I_2}\right)^{\beta_1} D_{11}^{\beta_1}.$$
(8.45)

Define $z = \frac{I_1}{I_2}$ and the function $f : [0,1] \to \mathbf{R}$ as

$$f(z) = z\beta_1 D_{10}^{\beta_1} - (\beta_1 - 1) D_{10}^{\beta_1} + z\beta_1 (D_{10} - D_{11}) D_{11}^{\beta_1 - 1} - z^{\beta_1} D_{11}^{\beta_1}.$$
(8.46)

Then we have for $z \in [0,1]$:

$$\frac{\partial f(z)}{\partial z} = \beta_1 D_{10}^{\beta_1} + \beta_1 (D_{10} - D_{11}) D_{11}^{\beta_1 - 1} - \beta_1 z^{\beta_1 - 1} D_{11}^{\beta_1}, \quad (8.47)$$

$$\frac{\partial^2 f(z)}{\partial z^2} = -\beta_1 (\beta_1 - 1) z^{\beta_1 - 2} D_{11}^{\beta_1} < 0. \quad (8.48)$$

Since

$$\begin{aligned}
\left.\frac{\partial f(z)}{\partial z}\right|_{z=1} &= \beta_1 D_{10}^{\beta_1} + \beta_1 (D_{10} - D_{11}) D_{11}^{\beta_1 - 1} - \beta_1 D_{11}^{\beta_1} \\
&= \beta_1 D_{10} \left(D_{10}^{\beta_1 - 1} - D_{11}^{\beta_1 - 1}\right) \\
&> 0, \quad (8.49)
\end{aligned}$$

we know that f is strictly concave and increasing on the interval $[0,1]$. Further, it holds that

$$f(0) = -(\beta_1 - 1) D_{10}^{\beta_1} < 0, \quad (8.50)$$
$$f(1) = D_{10}^{\beta_1} - \beta_1 (D_{10} - D_{11}) D_{11}^{\beta_1 - 1} - D_{11}^{\beta_1} > 0. \quad (8.51)$$

The last inequality follows from Lemma 8.1 with $u = D_{10}$, $v = D_{11}$, and $a = \beta_1$. The derived properties of f guarantee the existence and uniqueness of a z^* such that $f(z^*) = 0$. The threshold I_2^{**} equals $\frac{I_1}{z^*}$.   ☐

PROOF OF PROPOSITION 8.4 *From equations (8.7) and (8.15) we derive that $Y_{F_2} = Y_{L_1}$ if and only if $I_1 = I_1^*$, where*

$$I_1^* = I_2 \frac{D_{10}}{D_{11}}. \tag{8.52}$$

Thus the two cases of Proposition 8.4 apply. From equation (8.20) it follows that $I_1^* \leq I_2$. □

PROOF OF PROPOSITION 8.5 *The threshold Y_{L_2} can be derived in the same way as Y_{L_1}. Therefore*

$$Y_{L_2} = \frac{\beta_1}{\beta_1 - 1} \frac{(r - \mu) I_2}{D_{10}}. \tag{8.53}$$

Equations (8.7), (8.20), and (8.53) imply that $Y_{L_2} > Y_{F_1}$. □

B.   EQUILIBRIUM STRATEGIES

The equilibrium strategies for an asymmetric timing game are derived in the same way as in a symmetric timing game. For details we refer to Appendix A of Chapter 4, Simon, 1987a, and Simon, 1987b.

B.1   NEGATIVE EXTERNALITIES

For the case of negative externalities the equilibrium strategies are stated below. To derive these strategies the steps presented in Appendix A of Chapter 4 can be used.

1. If $I_2 \in [I_2^{**}, \infty)$ the equilibrium strategies of firms 1 and 2 in the negative externalities case for $t \geq 0$ are given by

$$G_1(t) = \begin{cases} 0 & \text{if } t < T_{L_1}, \\ 1 & \text{if } t \geq T_{L_1}, \end{cases}$$

$$\alpha_1(t) = \begin{cases} 0 & \text{if } t < T_{L_1}, \\ 1 & \text{if } t \geq T_{L_1}, \end{cases}$$

$$G_2(t) = \begin{cases} 0 & \text{if } t < T_{F_2}, \\ 1 & \text{if } t \geq T_{F_2}, \end{cases}$$

$$\alpha_2(t) = \begin{cases} 0 & \text{if } t < T_{F_2}, \\ 1 & \text{if } t \geq T_{F_2}. \end{cases}$$

2. If $I_2 \in (I_1, I_2^{**})$ four different cases should be analyzed.

(a) If $y \in (0, Y_{P_{21}}]$ the equilibrium strategies for $t \geq 0$ are given by

$$G_1(t) = \begin{cases} 0 & \text{if } t < T_{P_{21}}, \\ 1 & \text{if } t \geq T_{P_{21}}, \end{cases}$$

$$\alpha_1(t) = \begin{cases} 0 & \text{if } t < T_{P_{21}}, \\ 1 & \text{if } t \geq T_{P_{21}}, \end{cases}$$

$$G_2(t) = \begin{cases} 0 & \text{if } t < T_{F_2}, \\ 1 & \text{if } t \geq T_{F_2}, \end{cases}$$

$$\alpha_2(t) = \begin{cases} 0 & \text{if } t < T_{F_2}, \\ 1 & \text{if } t \geq T_{F_2}. \end{cases}$$

(b) If $y \in (Y_{P_{21}}, Y_{P_{22}})$ the equilibrium strategies are for $t \geq 0$ given by

$$G_1(t) = 1,$$
$$\alpha_1(t) = \frac{L_2(t) - F_2(t)}{L_2(t) - M_2(t)},$$
$$G_2(t) = 1,$$
$$\alpha_2(t) = \frac{L_1(t) - F_1(t)}{L_1(t) - M_1(t)}.$$

(c) If $y \in [Y_{P_{22}}, Y_{F_2})$ the equilibrium strategies for $t \geq 0$ are given by

$$G_1(t) = 1,$$
$$\alpha_1(t) = 1,$$
$$G_2(t) = \begin{cases} 0 & \text{if } t < T_{F_2}, \\ 1 & \text{if } t \geq T_{F_2}, \end{cases}$$
$$\alpha_2(t) = \begin{cases} 0 & \text{if } t < T_{F_2}, \\ 1 & \text{if } t \geq T_{F_2}. \end{cases}$$

(d) If $y \in [Y_{F_2}, \infty)$ the equilibrium strategies for $t \geq 0$ are equal to

$$G_1(t) = 1,$$
$$\alpha_1(t) = 1,$$
$$G_2(t) = 1,$$
$$\alpha_2(t) = 1.$$

B.2 POSITIVE EXTERNALITIES

For the case of positive externalities the equilibrium strategies are stated below. To derive these strategies the steps presented in Appendix A of Chapter 4 can be used.

1. If $I_1 \in (0, I_1^*)$ the equilibrium strategies for $t \geq 0$ are equal to

$$G_1(t) = \begin{cases} 0 & \text{if } t < T_{L_1}, \\ 1 & \text{if } t \geq T_{L_1}, \end{cases}$$

$$\alpha_1(t) = \begin{cases} 0 & \text{if } t < T_{L_1}, \\ 1 & \text{if } t \geq T_{L_1}, \end{cases}$$

$$G_2(t) = \begin{cases} 0 & \text{if } t < T_{F_2}, \\ 1 & \text{if } t \geq T_{F_2}, \end{cases}$$

$$\alpha_2(t) = \begin{cases} 0 & \text{if } t < T_{F_2}, \\ 1 & \text{if } t \geq T_{F_2}. \end{cases}$$

2. If $I_1 \in [I_1^*, I_2)$ the equilibrium strategies for $t \geq 0$ are equal to

$$G_i(t) = \begin{cases} 0 & \text{if } t < T_{F_2}, \\ 1 & \text{if } t \geq T_{F_2}, \end{cases}$$

$$\alpha_i(t) = \begin{cases} 0 & \text{if } t < T_{F_2}, \\ 1 & \text{if } t \geq T_{F_2}, \end{cases}$$

with $i \in \{1, 2\}$.

Chapter 9

TWO NEW TECHNOLOGIES

1.   INTRODUCTION

A firm that buys a new technology today faces the risk that a much better technology becomes available tomorrow. The fact that this can happen provides an incentive to delay the investment. To include this kind of mechanism, the chapter extends the models of Chapters 7 and 8 by incorporating an additional technology that becomes available at an unknown point of time in the future. This means that our model contains two different technologies that can be adopted, which are the currently available technology and a more efficient technology that becomes available at a future point of time. At the moment a firm invests, it enters the market, so, like in Chapter 8 we are considering a new market model. The reason is that we want to keep the model as simple as possible such that we are still able to point out the effects of adding an extra new technology. In this framework the possible invention of a more efficient technology raises the option value of waiting to invest in the current technology, but on the other hand the presence of a competitor may induce the firm to invest quickly, and thus forget about future technological progress.

The organization of the chapter is as follows. The model is presented in Section 2. After some preliminary analysis in Sections 3 and 4, the outcome of the game for different probabilities concerning the future appearance of the new technology is presented in Section 5. Section 6 collects the economic implications and Section 7 concludes.

217

2.   THE MODEL

We consider two identical, risk neutral and value maximizing firms that can make an investment expenditure $I\,(>0)$ to become active on a market. We denote the firms by i and j, with $i,j \in \{1,2\}$ and $i \neq j$. The firms discount future profits at rate $r\,(>0)$. At the beginning of the game, entering the market means producing with the existing technology 1. However, the decision to invest in technology 1 will be influenced by technological progress. Adopting technology 1 would have been a bad decision if a little later a much better technology becomes available. In our model technological progress is included as follows. At the stochastic time $T\,(>0)$ a new and better technology 2 becomes available for the firms. Time T is distributed according to an exponential distribution with mean $\frac{1}{\lambda}\,(>0)$, so that the arrival of technology 2 follows a Poisson process with parameter λ.

To be able to get analytical economic results we assume that firms can invest only once and that the investment costs of both technologies are equal. Concerning the profit flow it is assumed that it is stochastically evolving over time according to a geometric Brownian motion process. The profit flow of firm i at time $t\,(\geq 0)$ equals

$$\pi_i\,(t) = Y\,(t)\,D_{N_i N_j}, \qquad (9.1)$$

where N_k denotes the technology that firm $k\,(\in \{i,j\})$ is using. Hence, $N_k \in \{0,1,2\}$, where 0 means that the firm is not active. $Y\,(t)$ follows a geometric Brownian motion process

$$
\begin{aligned}
dY\,(t) &= \mu Y\,(t)\,dt + \sigma Y\,(t)\,d\omega\,(t), & (9.2)\\
Y\,(0) &= y, & (9.3)
\end{aligned}
$$

where $\mu\,(\in (0,r))$ is the drift parameter, $\sigma\,(>0)$ is the volatility parameter, $y\,(>0)$ is the starting value, and $d\omega\,(t)$ is an increment of a Wiener process. Thus $d\omega\,(t)$ is distributed according to a normal distribution with mean zero and variance dt. In the remainder of the chapter we omit the time dependence of $Y\,(t)$ whenever there is no confusion possible.

We make the following assumptions on the D's. First, a firm makes the highest amount of profits with a given technology if the other firm is not active (monopoly). It also holds that, given its own technology, profits are lowest when the other firm is a strong competitor, thus producing with the efficient technology 2. Second, given the technology of the competitor, the firm's profits are higher when it produces with the modern technology 2. In this way the following inequalities are obtained:

$$D_{20} > D_{21} > D_{22}$$
$$\vee \qquad \vee \qquad \vee \qquad (9.4)$$
$$D_{10} > D_{11} > D_{12}$$

Finally, since it is a new market model, firms do not earn anything as long as they have not adopted a technology. This implies that, for $N_k \in \{0, 1, 2\}$:

$$D_{0N_k} = 0. \qquad (9.5)$$

3. SECOND TECHNOLOGY BEING AVAILABLE

Three cases are possible when the second technology is already available. First, we consider the case where no firm has invested before time T, followed by the case where only the leader has invested before T. Finally, we give the payoff for the case that both firms have already invested before T.

3.1 NO INVESTMENT BEFORE TIME T

Since $t \geq T$, technology 2 is already available for adoption. This technology is more efficient than technology 1, and therefore the firms will never invest in technology 1. Hence, a game arises in which both firms consider entering a market by investing in *one* available technology, where the profit flow evolves stochastically over time. In fact, such a game is considered in Chapter 9 of Dixit and Pindyck, 1996, see also Chapter 7. In Chapter 7 it is shown that the expected value for each firm equals the follower value:

$$\Phi_{22}(Y) = \begin{cases} A_{22}Y^{\beta_1} & \text{if } Y < Y^F_{22}, \\ \frac{YD_{22}}{r-\mu} - I & \text{if } Y \geq Y^F_{22}, \end{cases} \qquad (9.6)$$

where

$$Y^F_{22} = \frac{\beta_1}{\beta_1 - 1} \frac{(r - \mu) I}{D_{22}}, \qquad (9.7)$$

$$A_{22} = \left(Y^F_{22}\right)^{-\beta_1} \left(\frac{Y^F_{22} D_{22}}{r - \mu} - I\right), \qquad (9.8)$$

$$\beta_1 = \frac{1}{2} - \frac{\mu}{\sigma^2} + \sqrt{\left(\frac{\mu}{\sigma^2} - \frac{1}{2}\right)^2 + \frac{2r}{\sigma^2}}. \qquad (9.9)$$

3.2   ONE INVESTMENT BEFORE TIME T

Here the leader has already invested in technology 1. Now the problem of the follower is in fact equal to that of a monopolist that considers entering a market where the profit flow equals YD_{21}. From the analysis of this standard investment problem (see, e.g., Dixit and Pindyck, 1996) it is obtained that the value of the follower equals

$$\Phi_{12}(Y) = \begin{cases} A_{12}Y^{\beta_1} & \text{if } Y < Y^F_{12}, \\ \frac{YD_{21}}{r-\mu} - I & \text{if } Y \geq Y^F_{12}, \end{cases} \tag{9.10}$$

where

$$Y^F_{12} = \frac{\beta_1}{\beta_1 - 1}\frac{(r-\mu)I}{D_{21}}, \tag{9.11}$$

$$A_{12} = \left(Y^F_{12}\right)^{-\beta_1}\left(\frac{Y^F_{12}D_{21}}{r-\mu} - I\right). \tag{9.12}$$

The value of the leader follows automatically:

$$\Lambda_{12}(Y) = \begin{cases} \frac{YD_{10}}{r-\mu} + B_{12}Y^{\beta_1} & \text{if } Y < Y^F_{12}, \\ \frac{YD_{12}}{r-\mu} & \text{if } Y \geq Y^F_{12}. \end{cases} \tag{9.13}$$

When $Y < Y^F_{12}$ the profit flow is too low for the follower to invest. Therefore the leader enjoys monopoly profits. If the leader receives these forever, the leader's total profits would equal $\frac{YD_{10}}{r-\mu}$. But it has to be taken into account that in the future Y could reach Y^F_{12} at a certain point of time. Then the follower will enter the market so that the leader's monopoly profits will be reduced. The term $B_{12}Y^{\beta_1}$ is the correction factor that incorporates this reduction into the firm's payoff for $Y < Y^F_{12}$. Therefore, the constant B_{12} is negative and, due to the fact that the leader's value function is continuous at Y^F_{12}, it can be derived that

$$B_{12} = \left(Y^F_{12}\right)^{1-\beta_1}\frac{D_{12} - D_{10}}{r-\mu}. \tag{9.14}$$

3.3   TWO INVESTMENTS BEFORE TIME T

The implication is that both firms have already invested in technology 1. Therefore, the value of each firm equals

$$\frac{YD_{11}}{r-\mu}. \tag{9.15}$$

4. SECOND TECHNOLOGY NOT BEING AVAILABLE

First, the follower's problem is analyzed, followed by the problem of the leader. Then we consider the joint mover payoff, and finally we determine the expected payoff in case both firms wait for technology 2.

4.1 FOLLOWER

First, we determine the follower's value if the follower waits for technology 2, while the leader has already invested in technology 1. Then we consider the case where the follower can also invest in technology 1, and determine the scenario under which investing in technology 1 can be optimal for the follower.

FOLLOWER WAITING FOR TECHNOLOGY 2

The value of the follower is denoted by $F_{12}(Y)$, and must satisfy the following Bellman equation

$$rF_{12}(Y) = \lim_{dt \downarrow 0} \frac{1}{dt} E\left[dF_{12}(Y) \right].$$ (9.16)

Itô's lemma (see Appendix A of Chapter 2) tells us that (for the definition of $\Phi_{12}(Y)$ see (9.10)):

$$E\left[dF_{12}(Y) \right] = (1 - \lambda dt)\left(\frac{\partial F_{12}(Y)}{\partial Y} \mu Y \, dt + \frac{\partial^2 F_{12}(Y)}{\partial Y^2} \frac{1}{2} \sigma^2 Y^2 dt \right)$$
$$+ \lambda dt \left(\Phi_{12}(Y) - F_{12}(Y) \right) + o(dt).$$ (9.17)

Substitution of (9.17) in (9.16) gives

$$\frac{\partial F_{12}(Y)}{\partial Y} \mu Y + \frac{\partial^2 F_{12}(Y)}{\partial Y^2} \frac{1}{2} \sigma^2 Y^2 - (r + \lambda) F_{12}(Y) + \lambda \Phi_{12}(Y) = 0.$$ (9.18)

Using the two possible expressions for $\Phi_{12}(Y)$ (see (9.10)), the solution of (9.18) equals

$$F_{12}(Y) = \begin{cases} \gamma_1 Y^{\beta_1^*} + A_{12} Y^{\beta_1} & \text{if } Y < Y_{12}^F, \\ \gamma_2 Y^{\beta_2^*} + \frac{\lambda}{r+\lambda-\mu} \frac{Y D_{21}}{r-\mu} - \frac{\lambda I}{r+\lambda} & \text{if } Y \geq Y_{12}^F, \end{cases}$$ (9.19)

where β_1^* (β_2^*) is the positive (negative) solution of

$$\frac{1}{2} \sigma^2 \beta^* (\beta^* - 1) + \mu \beta^* - (r + \lambda) = 0.$$ (9.20)

Expressions for γ_1 and γ_2 are found by solving the continuity and the differentiability conditions for F_{12} at $Y = Y_{12}^F$. This is done in Appendix A.1. It turns out that $\gamma_1 < 0$ and $\gamma_2 > 0$. In equation (9.19) we see that for $Y < Y_{12}^F$ the expected value of the follower consists of two parts. The second part equals the value of the option to adopt technology 2 (cf. equation (9.10)). The first part is a (negative) correction term, due to the fact that technology 2 is not available yet. Whenever Y is above the threshold Y_{12}^F the follower is going to adopt technology 2 at the moment that it becomes available. This last observation explains the last two terms of equation (9.19). The second term equals the expected present value of the profit flows generated from time T onwards:

$$
E\left[e^{-rT}\frac{Y(T)D_{21}}{r-\mu}\,\middle|\,Y(0)=Y\right]
$$

$$
= \frac{D_{21}}{r-\mu}E\left[e^{-rT}Y(T)\,\middle|\,Y(0)=Y\right]
$$

$$
= \frac{D_{21}}{r-\mu}\int_{t=0}^{\infty}\lambda e^{-\lambda t}e^{-rt}E\left[Y(t)\,\middle|\,Y(0)=Y\right]dt
$$

$$
= \frac{D_{21}}{r-\mu}\int_{t=0}^{\infty}\lambda e^{-\lambda t}e^{-rt}Ye^{\mu t}dt
$$

$$
= \frac{\lambda}{r+\lambda-\mu}\frac{YD_{21}}{r-\mu}. \tag{9.21}
$$

The third term is the expected present value of the investment cost that firm has to pay at time T in order to adopt technology 2:

$$
E\left[Ie^{-rT}\right]
$$

$$
= I\int_{t=0}^{\infty}\lambda e^{-\lambda t}e^{-rt}dt
$$

$$
= \frac{\lambda}{r+\lambda}I. \tag{9.22}
$$

Please note the difference between equations (9.21) and (9.22), i.e. the factors $\frac{\lambda}{r+\lambda-\mu}$ and $\frac{\lambda}{r+\lambda}$. In equation (9.21) the μ is subtracted from the denominator, in order to take into account the expected increase of Y.

If currently it holds that $Y(t) \geq Y_{12}^F$, it can still be the case that Y lies below the threshold Y_{12}^F at the time that the second technology arrives. Therefore, the correction term $\gamma_2 Y^{\beta_2^*}$, is added to the follower's value. This correction term is positive, since it reflects the fact that the firm is not committed to make an investment. Undertaking the investment would be suboptimal when Y is below Y_{12}^F at the moment the new technology is invented. Thus $\gamma_2 Y^{\beta_2^*}$ values flexibility. Notice that this correction factor vanishes when Y goes to infinity. This for the reason that the probability that $Y(T)$ is below Y_{12}^F goes to zero when Y goes to infinity.

FOLLOWER CONSIDERING TECHNOLOGY 1 TO BE INTERESTING

When Y increases, the opportunity costs of waiting rise. This could imply that, given that the probability that a more efficient technology is invented soon is sufficiently low, the follower is going to adopt technology 1 for large values of Y. Therefore, intuition suggests that, in case of λ sufficiently low, there exists a threshold Y_{11}^F such that the follower will wait with investing if $Y < Y_{11}^F$ and for $Y \geq Y_{11}^F$ the follower will adopt technology 1. Then the value of the follower is denoted by $F_{11}(Y)$ and equal to

$$
F_{11}(Y) = \begin{cases} \delta_1 Y^{\beta_1^*} + A_{12} Y^{\beta_1} & \text{if } Y \in \left[0, Y_{12}^F\right), \\ \delta_2 Y^{\beta_1^*} + \delta_3 Y^{\beta_2^*} + \frac{\lambda}{r+\lambda-\mu} \frac{YD_{21}}{r-\mu} - \frac{\lambda I}{r+\lambda} & \text{if } Y \in \left[Y_{12}^F, Y_{11}^F\right), \\ \frac{YD_{11}}{r-\mu} - I & \text{if } Y \in \left[Y_{11}^F, \infty\right). \end{cases}
$$

$$(9.23)$$

Equation (9.23) is derived by solving the follower's optimal stopping problem (see Appendix A.1). Solving the continuity and differentiability conditions for F_{11} at $Y = Y_{12}^F$ and the value matching and smooth pasting conditions for F_{11} at $Y = Y_{11}^F$ gives expressions for the constants δ_1, δ_2 and δ_3 (which can be found in Appendix A.1).

The term $\delta_1 Y^{\beta_1^*}$ consists of two parts. The first part, $(\delta_1 - \delta_2) Y^{\beta_1^*}$, is a correction term in the same fashion as $\gamma_1 Y^{\beta_1^*}$ and the second part, $\delta_2 Y^{\beta_1^*}$, is the value of the option to adopt technology 1. It turns out that the correction factor always dominates the option value and therefore $\delta_1 < 0$. The interpretation of $A_{12} Y^{\beta_1}$ is equal to the interpretation of the same factor in equation (9.19). The term $\delta_2 Y^{\beta_1^*}$ equals the option value of adopting technology 1, which implies that $\delta_2 > 0$. The correction factor $\delta_3 Y^{\beta_2^*}$ is exactly equal to $\gamma_2 Y^{\beta_2^*}$, thus $\delta_3 > 0$. Lemma 9.2 in Appendix B states the signs of the constants.

The following equation implicitly determines Y_{11}^F (cf. Appendix A.1):

$$
(\beta_1^* - \beta_2^*) \delta_3 \left(Y_{11}^F\right)^{\beta_2^*} + \frac{(\beta_1^* - 1)\lambda Y_{11}^F D_{21}}{(r+\lambda-\mu)(r-\mu)} - \frac{(\beta_1^* - 1)Y_{11}^F D_{11}}{r-\mu} + \frac{r\beta_1^* I}{r+\lambda} = 0. \quad (9.24)
$$

PROPOSITION 9.1 *The threshold Y_{11}^F has the following properties:*

1. Y_{11}^F *only exists if* $\lambda < \lambda_1^*$, *where*

$$\lambda_1^* = \frac{(r - \mu) D_{11}}{D_{21} - D_{11}}. \tag{9.25}$$

2. Y_{11}^F *approaches the follower's threshold for adopting technology 1 in a model without technology 2 (see Chapter 7) if* λ *approaches zero, i.e.*

$$\lim_{\lambda \downarrow 0} Y_{11}^F (\lambda) = \frac{\beta_1}{\beta_1 - 1} \frac{(r - \mu) I}{D_{11}}. \tag{9.26}$$

3. Y_{11}^F *approaches infinity if* λ *approaches* λ_1^*.

A proof of Proposition 9.1 can be found in Appendix B. It is intuitively clear that the threshold Y_{11}^F is rising with λ, but due to the complexity of expression (9.24) it was impossible to find an analytical proof for this statement. A larger λ implies that technology 2 is expected to arrive sooner and therefore it is in the follower's interest to postpone the adoption of technology 1. Hence, the threshold for adopting technology 1 will be set higher.

The follower postpones the adoption of technology 1 forever when Y_{11}^F approaches infinity. It is easy to verify that $\lim_{Y_{11}^F \to \infty} \delta_1 = \gamma_1$, $\lim_{Y_{11}^F \to \infty} \delta_2 = 0$ and $\delta_3 = \gamma_2$. This implies that equation (9.23) turns into equation (9.19) when Y_{11}^F goes to infinity.

4.2   LEADER

Here we consider the case where the leader invests in technology 1 (for the case where the leader invests in technology 2, see Subsection 3.1). Two scenarios are analyzed. In the first scenario the follower only considers investing in technology 2, while in the second scenario investing in technology 1 is an alternative for the follower.

FOLLOWER WAITING FOR TECHNOLOGY 2

When the follower waits for technology 2, the value of the leader equals

$$L_{12}(Y) = E \left[-I + \int_{t=0}^{T} Y(t) D_{10} e^{-rt} dt + e^{-rT} \Lambda_{12}(Y(T)) \middle| Y(0) = Y \right]. \tag{9.27}$$

This leads to the following expression for the leader curve (see (9.13), (9.14), and Appendix A.2)

$$L_{12}(Y) = \begin{cases} \varepsilon_1 Y^{\beta_1^*} + B_{12} Y^{\beta_1} + \frac{YD_{10}}{r-\mu} - I & \text{if } Y < Y_{12}^F, \\ \varepsilon_2 Y^{\beta_2^*} + \frac{YD_{10}}{r+\lambda-\mu} + \frac{\lambda}{r+\lambda-\mu} \frac{YD_{12}}{r-\mu} - I & \text{if } Y \geq Y_{12}^F. \end{cases}$$

(9.28)

Expressions for ε_1 and ε_2 are derived by solving the continuity and differentiability conditions for L_{12} at $Y = Y_{12}^F$, this is done in Appendix A.2. Lemma 9.4 in Appendix B states that ε_1 and ε_2 are both positive. The terms $\varepsilon_1 Y^{\beta_1^*}$ and $\varepsilon_2 Y^{\beta_2^*}$ correct for the fact that technology 2 has to arrive before the follower can adopt that technology and the leader's value becomes Λ_{12}. The longer it takes before technology 2 arrives, the longer the leader makes monopoly profits, i.e. the better for the leader. As in (9.13), $B_{12} Y^{\beta_1}$ stands for the option that Y exceeds Y_{12}^F, so that the follower will adopt technology 2, which ends the leader's monopoly profits. Consequently, as can be seen in (9.14), B_{12} is negative. The value $\varepsilon_2 Y^{\beta_2^*}$ equals the option that Y falls below Y_{12}^F. This is good for the leader because if $Y < Y_{12}^F$ the follower will not invest so that the leader keeps on having monopoly profits. This explains why ε_2 is positive.

FOLLOWER CONSIDERING TECHNOLOGY 1 TO BE INTERESTING

In this case the value of the leader is given by

$$L_{11}(Y) = \begin{cases} \phi_1 Y^{\beta_1^*} + B_{12} Y^{\beta_1} + \frac{YD_{10}}{r-\mu} - I & \text{if } Y \in [0, Y_{12}^F), \\ \phi_2 Y^{\beta_1^*} + \phi_3 Y^{\beta_2^*} + \frac{YD_{10}}{r+\lambda-\mu} & \\ + \frac{\lambda}{r+\lambda-\mu} \frac{YD_{12}}{r-\mu} - I & \text{if } Y \in [Y_{12}^F, Y_{11}^F), \\ \frac{YD_{11}}{r-\mu} - I & \text{if } Y \in [Y_{11}^F, \infty). \end{cases}$$

(9.29)

The derivation of equation (9.29) and expressions for ϕ_1, ϕ_2 and ϕ_3 can be found in Appendix A.2. The signs of ϕ_1 and ϕ_3 are equal to the signs of ε_1 and ε_2 in (9.28), respectively (see Lemma 9.5 in Appendix B).

The constant ϕ_2 values the possibility that Y rises above Y_{11}^F before technology 2 arrives. On the one hand that event is good for the leader, since the follower adopts technology 1 and not technology 2. On the other hand it is bad for the leader, because it no longer has a monopoly position. The following proposition states under which conditions ϕ_2 is negative or positive, i.e. which argument dominates the other. The proof is given in Appendix B.

PROPOSITION 9.2 *A sufficient condition for the constant ϕ_2 to be non-positive is*

$$\frac{D_{21}}{D_{11}} \geq \frac{D_{12} - D_{10}}{D_{11} - D_{10}}. \tag{9.30}$$

If equation (9.30) does not hold, the sign of ϕ_2 can go both ways.

Equation (9.30) states that the relative profit gain the follower can make by adopting technology 2 is larger than the relative profit loss that the leader faces when the follower adopts technology 2. Inequality (9.30) is most likely to hold when the leader is almost indifferent concerning the technology the follower switches to. In that case it is not good for the leader if the follower switches to 1 immediately rather than waiting for 2. Consequently ϕ_2 is negative which is confirmed by Proposition 9.2.

4.3   JOINT INVESTMENT

The expected value of each firm if both firms adopt technology 1 together is given by

$$M_{11}(Y) = \frac{Y D_{11}}{r - \mu} - I. \tag{9.31}$$

4.4   WAITING CURVE

The waiting curve (see also Chapter 6) gives the expected value if both firms wait with investing until technology 2 arrives. The waiting curve equals

$$
\begin{aligned}
W(Y) &= E\left[e^{-rT}\Phi_{22}\left(Y(T)\right)\middle| Y(0) = Y\right] \\
&= \begin{cases}
\eta_1 Y^{\beta_1^*} + A_{22} Y^{\beta_1} & \text{if } Y < Y_{22}^F, \\
\eta_2 Y^{\beta_2^*} + \frac{\lambda Y D_{22}}{(r+\lambda-\mu)(r-\mu)} - \frac{\lambda I}{r+\lambda} & \text{if } Y \geq Y_{22}^F.
\end{cases} \tag{9.32}
\end{aligned}
$$

For a derivation we refer to Appendix A.3, there we also present expressions for η_1 and η_2. The constant η_1 is negative and the constant η_2 is positive. These constants have the same economic interpretations as γ_1 and γ_2, respectively.

PROPOSITION 9.3 *It always holds that $F_{12}(Y) > W(Y)$.*

This proposition is proved in Appendix B and is a direct result of the new market assumption. The follower starts making profits after its investment and from the follower's point of view it is best that the leader adopts technology 1.

5. EQUILIBRIA

In this section firm roles are endogenous which means that it is not determined beforehand which firm will be the first investor. We describe the possible equilibria of the technology adoption game before the arrival of technology 2. It turns out that the type of the equilibria is completely determined by λ. In the following theorem we describe this relationship.

THEOREM 9.1 *There are three regions for* λ.

1. *If* $\lambda \in [0, \lambda_2^*)$ *the equilibrium is of the preemption type.*

2. *If* $\lambda \in [\lambda_2^*, \lambda_3^*)$ *the equilibrium is of the attrition type.*

3. *If* $\lambda \in [\lambda_3^*, \infty)$ *both firms wait with investing until technology 2 arrives.*

The critical λ *levels are equal to*

$$\lambda_2^* = \frac{(r - \mu)\, D_{10}}{D_{21} - D_{12}}, \tag{9.33}$$

$$\lambda_3^* = \frac{(r - \mu)\, D_{10}}{D_{22} - D_{12}}. \tag{9.34}$$

The first λ region is split up into two λ regions: $[0, \lambda_1^*)$ and $[\lambda_1^*, \lambda_2^*)$, where λ_1^* is given by (9.25). Note that equation (9.30) ensures that $\lambda_1^* \leq \lambda_2^*$. In case equation (9.30) does not hold, the second region for λ does not exist. In each of the following four subsections one of the regions for λ is analyzed and the equilibria are characterized. In the remainder of this section Theorem 9.1 is implicitly proved. The propositions in this section are proved in Appendix B. We do not prove the theorems. Interested readers are referred to Appendix A of Chapter 4 and Chapter 6 where the equilibrium concepts are presented.

5.1 CASE 1

In the first case we have $\lambda \in [0, \lambda_1^*)$. From the analysis of the previous section we know that in this region the follower is going to adopt technology 1 for Y large enough. This implies that in the equilibrium analysis the leader curve is given by equation (9.29), the follower curve by (9.23), the joint investment curve by (9.31), and the waiting curve by (9.32). The following proposition states that there exists a preemption threshold in this region.

PROPOSITION 9.4 *Let* $\lambda \in [0, \lambda_1^*)$. *Then there exists a unique* $Y_{11}^P \in (0, Y_{11}^F)$ *such that*

$$L_{11}\left(Y_{11}^P\right) = F_{11}\left(Y_{11}^P\right). \tag{9.35}$$

Define T_{11}^P and T_{11}^F as follows: $T_{11}^P = \inf\left(t \mid Y(t) \geq Y_{11}^P\right)$ and $T_{11}^F = \inf\left(t \mid Y(t) \geq Y_{11}^F\right)$. Propositions 9.3 and 9.4 imply that the leader curve exceeds the waiting curve for some Y. From Chapter 6 it follows that the equilibria of this game with waiting curve are equal to equilibria of the game without waiting curve. This means that in analyzing the game the future arrival of technology 2 can be ignored for the moment (of course, if, despite the low probability, technology 2 arrives before one of the firms has invested in technology 1, the outcome must be reconsidered). Hence, a game must be considered where two firms have to determine their optimal timing concerning the investment in a given technology. This is in fact the game described in Chapter 9 of Dixit and Pindyck, 1996, see also Chapter 7. Here we repeat the most important aspects.

THEOREM 9.2 *Consider the game with* $y \leq Y_{11}^P$. *It holds that in equilibrium the leader adopts technology 1 at time* T_{11}^P *and the follower adopts technology 1 at time* T_{11}^F.

Of course, Theorem 9.2 is conditional on the fact that technology 2 does not arrive before time T_{11}^F. Further we should remark that if $Y_{11}^P < y < Y_{11}^F$ there exists a positive probability that the firms invest simultaneously at time 0 (cf. Appendix A.2 of Chapter 4 and Chapter 7). In equilibrium the expected value of each firm equals the follower value. Figure 9.1 graphically shows the curves in this case.

The investment opportunity is worthless for Y equal to zero. Therefore, at $Y = 0$ the leader (L) and joint investment (M) value equal minus the investment cost and the follower (F) value equals zero. The further shape of the curves L, F, M, and W can be derived from (9.23), (9.29), (9.31), (9.32), and (9.35).

With Figure 9.1 the preemption mechanism can be clearly explained. Consider the game with $Y(0) \leq Y_{11}^P$. Assume that both firms pass Y_{11}^P without investing and the current value of Y, say $Y(t)$, exceeds Y_{11}^P. Then for one of the firms it is optimal to invest at time t, since the L-curve lies above the F-curve, implying that investing first gives a higher payoff than investing second. The other firm knows this and will try to preempt its competitor by investing at time $t - \epsilon$, since it knows that the other firm would like to be the first to invest at time t. But then the other firm will try to preempt at time $t - 2\epsilon$. It is clear that this process of preemption stops at Y_{11}^P, since for $Y < Y_{11}^P$ it holds that $F(Y) > L(Y)$ so that there are no incentives to invest first.

The following proposition gives an expression for the probability that technology 2 arrives after a certain threshold is hit. The proof of the proposition is given in Appendix B.

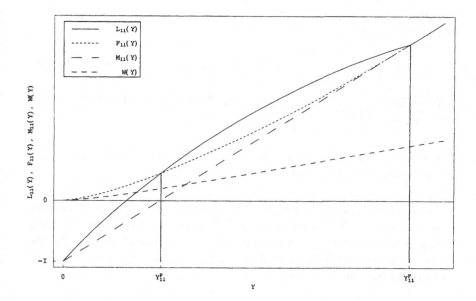

Figure 9.1.   Case 1: $\lambda \in [0, \lambda_1^*)$.

PROPOSITION 9.5 *Let* $T_S = \inf (t | Y (t) \geq S)$. *At time* $t = 0$ *the probability that the geometric Brownian motion hits the threshold S before the second technology arrives, i.e.* $\Pr (T_S < T)$, *is given by*

$$\Pr (T_S < T) = \begin{cases} \left(\frac{y}{S}\right)^{\widehat{\beta}_1} & \text{if } y < S, \\ 1 & \text{if } y \geq S, \end{cases} \tag{9.36}$$

where

$$\widehat{\beta}_1 = \frac{1}{2} - \frac{\mu}{\sigma^2} + \sqrt{\left(\frac{\mu}{\sigma^2} - \frac{1}{2}\right)^2 + \frac{2\lambda}{\sigma^2}}. \tag{9.37}$$

From Proposition 9.5 we derive that the probability that technology 1 is adopted by the leader (follower) decreases with λ. An increase of λ leads to both a higher threshold and a higher $\widehat{\beta}_1$.

5.2   CASE 2

In the second case it holds that $\lambda \in [\lambda_1^*, \lambda_2^*)$. Here the probability that technology 2 arrives soon is that high that the follower is going to wait for technology 2. As in the previous case there exists a preemption threshold.

PROPOSITION 9.6 *Let* $\lambda \in [\lambda_1^*, \lambda_2^*)$. *Then there exists a unique* $Y_{12}^P \in (0, \infty)$ *such that*

$$L_{12} \left(Y_{12}^P \right) = F_{12} \left(Y_{12}^P \right). \tag{9.38}$$

We define T_{12}^P in the same fashion as T_{11}^P: $T_{12}^P = \inf \left(t \,|\, Y(t) \geq Y_{12}^P \right)$. Furthermore we define $T_{12}^F = \inf \left(t \geq T \,|\, Y(t) \geq Y_{12}^F \right)$.

THEOREM 9.3 *In equilibrium the leader adopts technology 1 at time* T_{12}^P *and the follower adopts technology 2 at time* T_{12}^F.

As above the leader's adoption of technology 1 is conditional on technology 2 not arriving before time T_{12}^P. If initially Y is above Y_{12}^P then with positive probability both firms adopt technology 1 at time 0. The expected value of each firm equals the follower value. The curves for this case are plotted in Figure 9.2.

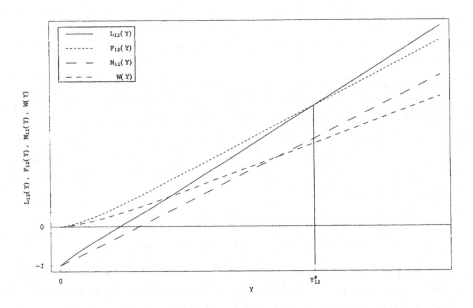

Figure 9.2.   Case 2: $\lambda \in [\lambda_1^*, \lambda_2^*)$.

From the fact that Y_{12}^P is rising in λ and Proposition 9.5 it can be concluded that the probability that the leader adopts technology 1 decreases with λ.

5.3 CASE 3

The third case is characterized by the fact that $\lambda \in [\lambda_2^*, \lambda_3^*)$. Here the probability that technology 2 arrives is even higher than in case 2, where it was already high enough for the follower to wait for technology 2. This implies that also in this case the follower is going to wait for technology 2. In this region there does not exist a preemption threshold, i.e. the follower curve is situated above the leader curve for each Y. This implies that the game without waiting curve is an attrition game.

PROPOSITION 9.7 *Let $\lambda \in [\lambda_2^*, \lambda_3^*)$. Then there exists a unique $Y_{12}^L \in (0, \infty)$ such that*

$$L_{12}\left(Y_{12}^L\right) = W\left(Y_{12}^L\right). \tag{9.39}$$

The following theorem describes the equilibrium conditional on technology 2 not arriving before time $T_{12}^L = \inf\left(t \,|\, Y(t) \geq Y_{12}^L\right)$.

THEOREM 9.4 *In equilibrium the leader adopts technology 1 at time T_{12}^L and the follower adopts technology 2 at time T_{12}^F.*

The curves for the different payoffs in this game are depicted in Figure 9.3. The leader curve shows the expected payoff as function of Y for a firm that invests in technology 1 immediately. This firm knows that its competitor will invest in technology 2 as soon as it becomes available and $Y > Y_{12}^F$. The leader has the advantage of monopoly profits until the time that the follower invests in technology 2, but the disadvantage of producing with a less efficient technology after this date. On the other hand the waiting curve shows the expected payoff if both firms wait for technology 2 to arrive. As long as the waiting curve lies above the leader curve, investing now in technology 1 is not a sensible option. Therefore, the attrition game starts at time T_{12}^L.

In the attrition game the follower curve is situated above the leader curve and the leader curve above the joint investment curve for all positive Y. This implies that there does not exist a symmetric equilibrium for this attrition game (cf. Appendix A.3 of Chapter 4). There are two asymmetric equilibria, which are summarized in Theorem 9.4 (each firm can either be leader or follower). For simplicity reasons we assume that each equilibrium occurs with probability one half.

Proposition 9.5 together with the fact that Y_{12}^L is increasing with λ imply that the probability that the leader adopts technology 1 is decreasing with λ.

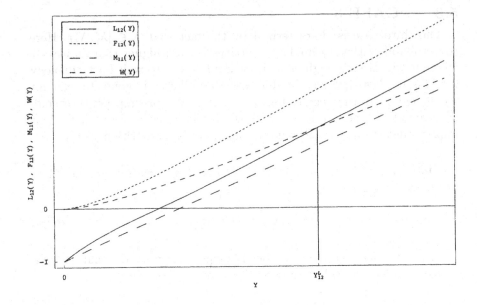

Figure 9.3.   Case 3: $\lambda \in [\lambda_2^*, \lambda_3^*)$.

5.4   CASE 4

In the fourth case ($\lambda \in [\lambda_3^*, \infty)$) the probability that technology 2 will be invented soon is that high that both firms wait with investing until technology 2 arrives. This is reflected by the fact that the waiting curve exceeds the leader curve for all Y in this region. Figure 9.4 shows the curves in this case.

At the moment that technology 2 arrives, a game starts where both firms consider entering a market by investing in *one* available technology (the presence of technology 1 can be ignored since it is less efficient), while the profit flow follows a geometric Brownian motion process. Hence, like in case 1, the framework of Chapter 9 of Dixit and Pindyck, 1996 again applies. The difference is that in case 1 the Dixit and Pindyck game has to deal with investment in technology 1, while here the investment in technology 2 must be considered.

6.   ECONOMIC ANALYSIS

The Poisson parameter λ is the key parameter for the results. Waiting for the new technology is better when the probability that this new technology becomes available soon, is high enough. If this probability is low enough both firms only consider when to invest in the current

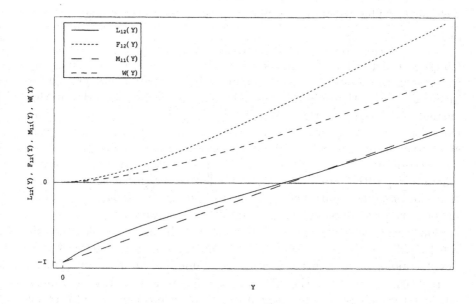

Figure 9.4.   Case 4: $\lambda \in [\lambda_3^*, \infty)$.

technology, while ignoring the new one. In this case the usual preemption game arises (see Fudenberg and Tirole, 1985 and Chapter 9 of Dixit and Pindyck, 1996 for its stochastic counterpart).

If the probability that technology 2 becomes available soon is not too small, i.e. the Poisson parameter exceeds λ_1^* (cf. (9.25)), then the game is still a preemption game, so that each firm tries to be the first investor. However, the firm that will invest second is better off by waiting for the new technology rather than investing in the current one.

If λ is again a bit larger such that it exceeds λ_2^* (see (9.33)), the preemption game turns into an attrition game. Like in the previous case, the first investor chooses the current technology and the second investor will wait for the new technology, but the difference is that the payoff of the second investor is higher here. Hence, neither firm would like to be the first investor, but if they both keep on waiting, their payoff will be even less than the payoff of the one that decides immediately to invest first. According to Appendix A.3 of Chapter 4 a unique asymmetric equilibrium exists where the adoption timings are dispersed.

If λ exceeds λ_3^*, given by (9.34), then the probability that technology 2 arrives soon is that large that both firms will wait for this new technology. The possibility to invest in the current technology will be ignored.

It is clear that for $\lambda = 0$ the model exactly equals the one treated in Chapter 9 of Dixit and Pindyck, 1996. Here there is no technological progress in the sense that the probability that a new technology will be invented is zero. Hence both firms only need to consider investing in the current technology, so that the problem boils down to the determination of the optimal point in time that a firm must enter a market with stochastic profit flow, while taking into account the behavior of an identical competitor. The resulting game is a preemption game, like the one where λ is positive but below λ_1^*. It holds that Y_{11}^P increases with λ so that the possible occurrence of a new technology will delay investment in the current technology, which is intuitively plausible.

Comparing the case for $\lambda = 0$ (model with one technology) with $\lambda \in (\lambda_2^*, \lambda_3^*)$ shows that taking into account the possible occurrence of a new technology could turn a preemption game into an attrition game.

To learn more about the effects of the future availability of a more efficient technology on the optimal timing of investment, we also carry out comparative statics analysis on the other parameter values. Let us first consider the effect of revenue volatility which is measured by σ. The general prediction of the real options literature is that a higher level of uncertainty increases the threshold level and therefore will have a negative effect on investment. In our model an increased threshold level implies that the investment in technology 1 will be delayed. Therefore, the probability that technology 2 arrives before the investment is undertaken, increases. Hence, the conclusion is that increased revenue uncertainty induces a higher probability that the new technology will be adopted instead of the current technology.

Next, consider the expected growth of the market reflected by the parameter μ. An increase of μ reduces the values of λ_i^*, with $i \in \{1, 2, 3\}$. In general this means that the probability increases that the firm will delay or totally refrain from investing in the current technology. The reason is that in case of a fast growing market the firm will exploit this growth as much as possible by using the more efficient new technology. The firm is more willing to wait for this technology to be invented.

The effect of the discount rate is completely opposite to the effect of the expected market growth rate. A higher discount rate implies that immediate profits are more important to the firm. Therefore the firm prefers investing in the current technology rather than waiting for the new one.

Finally, consider the effects of the several profit flows. First, notice that λ_1^* increases with D_{11} and decreases with D_{21}. This can be explained by the fact that the second investor is more willing to produce with the first technology if D_{11} is large, while it likes to wait for the new technology to arrive if D_{21} is large.

Second, λ_2^* increases with D_{10} and D_{12}, while it decreases with D_{21}. This implies that λ_2^* is larger if the payoff of the strategy "adopt technology 1 immediately" is higher relative to the payoff of the strategy "wait for technology 2 to arrive and adopt it then". Note that if the latter strategy gives the highest payoff, the game is an attrition game, which occurs for $\lambda \in (\lambda_2^*, \lambda_3^*)$.

Third, λ_3^* increases with D_{10} and D_{12}, while it decreases with D_{22}. Hence, if a high profit is reached when both firms produce with the new technology, compared to the strategy "invest in technology 1 immediately and have some monopoly profits before technology 2 arrives", both firms will wait for the second technology to arrive. This in fact happens for $\lambda > \lambda_3^*$.

7. CONCLUSIONS

The optimal investment timing is governed to a large extent by the magnitude of the probability that the new technology becomes available within a given period of time. We found that, indeed, the possible occurrence of a new technology will delay investment in the current technology. Compared to the case where technological progress is not included (for example Chapters 7 and 8), taking into account the possible occurrence of a new technology could turn a preemption game into an attrition game, which is a game where the second mover gets the highest payoff. This could happen when the first mover invests in the current technology, while the second mover waits for the new technology to arrive and invests then in it, and can be explained as follows. Compared to the strategy of its competitor, the benefits of the first investor are the monopoly profits gained during the period that starts at the moment of investment by the first investor and lasts until the moment that the second mover invests. However, these monopoly profits can be more than offset by the efficiency gain the second investor enjoys due to producing with a more efficient technology, which takes place after both firms have invested.

From the theory of real options it is known that the option value of waiting with investment increases with revenue uncertainty. For our model this implies that increased uncertainty delays adoption of the current technology, so that the probability that the new technology is invented before the investment in the current technology has taken place increases. This leads to the conclusion that increased revenue uncertainty induces a higher probability that the new technology will be adopted instead of the current technology. Hence, uncertainty raises the technological level within firms. Another result that is worth mentioning here is, that in a faster growing market a firm is more inclined to wait for a more efficient technology to arrive.

Appendices
A.   DERIVATION OF VALUE FUNCTIONS
A.1   FOLLOWER
FOLLOWER WAITING FOR TECHNOLOGY 2

Solving the continuity and differentiability conditions for $F_{12}(Y)$ at $Y = Y_{12}^F$ gives

$$\gamma_1 = \frac{\left(Y_{12}^F\right)^{-\beta_1^*} I\left(r(r-\mu)\beta_2^* + (r-\mu\beta_1)\lambda\beta_2^* - (r-\mu)(r+\lambda)\beta_1\right)}{(r+\lambda)(r+\lambda-\mu)(\beta_1-1)\left(\beta_1^*-\beta_2^*\right)}, \qquad (9.40)$$

$$\gamma_2 = \frac{\left(Y_{12}^F\right)^{-\beta_2^*} I\left(r(r-\mu)\beta_1^* + (r-\mu\beta_1)\lambda\beta_1^* - (r-\mu)(r+\lambda)\beta_1\right)}{(r+\lambda)(r+\lambda-\mu)(\beta_1-1)\left(\beta_1^*-\beta_2^*\right)}. \qquad (9.41)$$

A direct result of Lemma 9.1 (see Appendix B) is that $\gamma_1 < 0$ and $\gamma_2 > 0$.

FOLLOWER CONSIDERING TECHNOLOGY 1 TO BE INTERESTING

The follower solves the optimal stopping problem, in which stopping means adopting technology 1. Therefore the expected value of the follower for $Y \geq Y_{11}^F$ equals

$$F_{11}(Y) = \frac{Y D_{11}}{r-\mu} - I. \qquad (9.42)$$

In the continuation region waiting is the optimal strategy and the following Bellman equation must be satisfied

$$r F_{11}(Y) = \lim_{dt \downarrow 0} \frac{1}{dt} E\left[dF_{11}(Y)\right]. \qquad (9.43)$$

Expanding the right-hand-side of (9.43) with Itô's lemma and rewriting gives

$$\frac{\partial F_{11}(Y)}{\partial Y}\mu Y + \frac{\partial^2 F_{11}(Y)}{\partial Y^2}\frac{1}{2}\sigma^2 Y^2 - (r+\lambda) F_{11}(Y) + \lambda\Phi_{12}(Y) = 0. \qquad (9.44)$$

Using (9.10) and the boundary condition $F_{11}(0) = 0$ gives

$$
F_{11}(Y) = \begin{cases} \delta_1 Y^{\beta_1^*} + A_{12} Y^{\beta_1} & \text{if } Y < Y_{12}^F, \\ \delta_2 Y^{\beta_1^*} + \delta_3 Y^{\beta_2^*} + \frac{\lambda Y D_{21}}{(r+\lambda-\mu)(r-\mu)} - \frac{\lambda I}{r+\lambda} & \text{if } Y \geq Y_{12}^F. \end{cases}
$$

(9.45)

Combining (9.42) and (9.43) gives equation (9.23).

Expressions for δ_1, δ_2, δ_3 and Y_{11}^F are found by simultaneously solving the continuity and differentiability conditions for F_{11} at Y_{12}^F and the value matching and smooth pasting conditions for F_{11} at Y_{11}^F. It turns out that it is not possible to get a closed form solution for Y_{11}^F. The threshold Y_{11}^F is implicitly determined by equation (9.24). The constants are equal to

$$
\delta_1 = \delta_2 + \left(Y_{12}^F\right)^{\beta_2^* - \beta_1^*} \frac{\beta_2^* \gamma_2}{\beta_1^*} - \left(Y_{12}^F\right)^{1-\beta_1^*} \frac{D_{21}}{\beta_1^*(r+\lambda-\mu)},
$$

(9.46)

$$
\delta_2 = \left(Y_{11}^F\right)^{\beta_2^* - \beta_1^*} \frac{(1-\beta_2^*)\gamma_2}{\beta_1^* - 1} + \left(Y_{11}^F\right)^{-\beta_1^*} \frac{rI}{(\beta_1^* - 1)(r+\lambda)},
$$

(9.47)

$$
\delta_3 = \gamma_2.
$$

(9.48)

Lemma 9.2 in Appendix B states that $\delta_1 < 0$, $\delta_2 > 0$, and $\delta_3 > 0$.

A.2 LEADER
FOLLOWER WAITING FOR TECHNOLOGY 2

In order to derive an expression for equation (9.27), define

$$
h(Y) = E\left[\int_{t=0}^{T} Y(t) D_{10} e^{-rt} dt + e^{-rT} \Lambda_{12}(Y(T)) \,\middle|\, Y(0) = Y\right].
$$

(9.49)

Then $h(Y)$ must satisfy the following Bellman equation

$$
rh(Y) = Y D_{10} + \lim_{dt \downarrow 0} \frac{1}{dt} E[dh(Y)].
$$

(9.50)

Applying Itô's lemma gives

$$
E[dh(Y)] = (1 - \lambda dt)\left(\frac{\partial h(Y)}{\partial Y} \mu Y dt + \frac{\partial^2 h(Y)}{\partial Y^2} \frac{1}{2} \sigma^2 Y^2 dt\right)
$$
$$
+ \lambda dt \left(\Lambda_{12}(Y) - h(Y)\right) + o(dt).
$$

(9.51)

Substitution of (9.51) in (9.50) gives

$$
\frac{\partial h(Y)}{\partial Y} \mu Y + \frac{\partial^2 h(Y)}{\partial Y^2} \frac{1}{2} \sigma^2 Y^2 - (r+\lambda) h(Y) + \lambda \Lambda_{12}(Y) + Y D_{10} = 0.
$$

(9.52)

Substitution of (9.13) in (9.52) and solving that differential equation gives

$$h\left(Y\right) = \begin{cases} \varepsilon_1 Y^{\beta_1^*} + \tau_1 Y^{\beta_2^*} + B_{12} Y^{\beta_1} + \frac{Y D_{10}}{r-\mu} & \text{if } Y < Y_{12}^F, \\ \tau_2 Y^{\beta_1^*} + \varepsilon_2 Y^{\beta_2^*} + \frac{Y D_{10}}{r+\lambda-\mu} + \frac{\lambda}{r+\lambda-\mu} \frac{Y D_{12}}{r-\mu} & \text{if } Y \geq Y_{12}^F. \end{cases}$$

(9.53)

The boundary condition at $Y = 0$ and the condition that rules out speculative bubbles (see p. 181 of Dixit and Pindyck, 1996),

$$h\left(0\right) = 0, \tag{9.54}$$

$$\lim_{Y \to \infty} \frac{h\left(Y\right)}{Y} = \frac{D_{10}}{r+\lambda-\mu} + \frac{\lambda}{r+\lambda-\mu} \frac{D_{12}}{r-\mu}, \tag{9.55}$$

imply that $\tau_1 = 0$ and $\tau_2 = 0$.

Solving the continuity and differentiability conditions for L_{12} at $Y = Y_{12}^F$ gives

$$\varepsilon_1 = \frac{\left(Y_{12}^F\right)^{1-\beta_1^*}\left((r-\mu)(\beta_1-\beta_2^*)+\lambda(\beta_1-1)\right)(D_{10}-D_{12})}{(r+\lambda-\mu)(r-\mu)\left(\beta_1^*-\beta_2^*\right)}, \tag{9.56}$$

$$\varepsilon_2 = \frac{\left(Y_{12}^F\right)^{1-\beta_2^*}\left((r-\mu)(\beta_1-\beta_1^*)+\lambda(\beta_1-1)\right)(D_{10}-D_{12})}{(r+\lambda-\mu)(r-\mu)\left(\beta_1^*-\beta_2^*\right)}. \tag{9.57}$$

According to Lemma 9.4 ε_1 and ε_2 are both positive (see Appendix B).

FOLLOWER CONSIDERING TECHNOLOGY 1 TO BE INTERESTING

If $Y \geq Y_{11}^F$ the value function of the leader is given by

$$L_{11}\left(Y\right) = \frac{Y D_{11}}{r - \mu} - I. \tag{9.58}$$

Next we derive the value function of the leader for $Y < Y_{11}^F$. The value of the leader equals

$$L_{11}\left(Y\right) = E \left[-I + \int_{t=0}^{\min\left(T,T_{11}^F\right)} Y\left(t\right) D_{10} e^{-rt} dt \right.$$

(9.59)

$$+ 1_{\{T \leq T_{11}^F\}} e^{-rT} \Lambda_{12}\left(Y\left(T\right)\right)$$

$$+ \left. \int_{t=\min\left(T,T_{11}^F\right)}^{\infty} 1_{\{T_{11}^F < T\}} Y\left(t\right) D_{11} e^{-rt} dt \,\middle|\, Y\left(0\right) = Y \right].$$

Define

$$f(Y) = E\left[\int_{t=0}^{\min(T,T_{11}^F)} Y(t)D_{10}e^{-rt}dt \,\Big|\, Y(0) = Y\right]. \qquad (9.60)$$

The function f must satisfy the following Bellman equation for $Y < Y_{11}^F$:

$$rf(Y) = YD_{10} + \lim_{dt\downarrow 0}\frac{1}{dt}E\left[df(Y)\right]. \qquad (9.61)$$

Itô's lemma gives

$$E\left[df(Y)\right] = \lambda dt\left(0 - f(Y)\right) + (1 - \lambda dt)\left(\frac{\partial f(Y)}{\partial Y}\mu Y dt\right.$$
$$\left. + \frac{\partial^2 f(Y)}{\partial Y^2}\frac{1}{2}\sigma^2 Y^2 dt\right) + o(dt). \qquad (9.62)$$

Thus

$$\mu Y\frac{\partial f(Y)}{\partial Y} + \frac{1}{2}\sigma^2 Y^2\frac{\partial^2 f(Y)}{\partial Y^2} - (r + \lambda)f(Y) + YD_{10} = 0. \qquad (9.63)$$

The solution of this differential equation is given by

$$f(Y) = v_1 Y^{\beta_1^*} + v_2 Y^{\beta_2^*} + \frac{YD_{10}}{r + \lambda - \mu}. \qquad (9.64)$$

Using the boundary conditions $f(0) = 0$ and $f\left(Y_{11}^F\right) = 0$ the values for the constants are found:

$$v_1 = -\left(Y_{11}^F\right)^{1-\beta_1^*}\frac{D_{10}}{r + \lambda - \mu}, \qquad (9.65)$$

$$v_2 = 0. \qquad (9.66)$$

Next define

$$g(Y) = E\left[1_{\{T\leq T_{11}^F\}}e^{-rT}\Lambda_{12}(Y(T))\right. \qquad (9.67)$$

$$\left. + \int_{t=\min(T,T_{11}^F)}^{\infty} 1_{\{T_{11}^F<T\}}Y(t)D_{11}e^{-rt}dt \,\Big|\, Y(0) = Y\right].$$

The function g must satisfy the following Bellman equation

$$rg(Y) = \lim_{dt\downarrow 0}\frac{1}{dt}\left(\lambda dt\left(\Lambda_{12}(Y) - g(Y)\right)\right.$$

$$\left. + (1 - \lambda dt)\left(\frac{\partial g(Y)}{\partial Y}\mu Y dt + \frac{\partial^2 g(Y)}{\partial Y^2}\frac{1}{2}\sigma^2 Y^2 dt + o(dt)\right)\right),$$

$$(9.68)$$

leading to

$$\mu Y \frac{\partial g\left(Y\right)}{\partial Y} + \frac{1}{2}\sigma^2 Y^2 \frac{\partial^2 g\left(Y\right)}{\partial Y^2} - \left(r + \lambda\right) g\left(Y\right) + \lambda \Lambda_{12}\left(Y\right) = 0. \quad (9.69)$$

The solution of (9.69) is given by

$$g\left(Y\right) = \begin{cases} \kappa_1 Y^{\beta_1^*} + \kappa_2 Y^{\beta_2^*} + B_{12} Y^{\beta_1} + \frac{\lambda}{r+\lambda-\mu}\frac{YD_{10}}{r-\mu} & \text{if } Y < Y_{12}^F, \\ \kappa_3 Y^{\beta_1^*} + \kappa_4 Y^{\beta_2^*} + \frac{\lambda}{r+\lambda-\mu}\frac{YD_{12}}{r-\mu} & \text{if } Y \geq Y_{12}^F. \end{cases}$$
$$(9.70)$$

Due to the boundary condition $g\left(0\right) = 0$ we know that $\kappa_2 = 0$. The constants κ_1, κ_3, and κ_4 are found by simultaneously solving the continuity and differentiability condition at $Y = Y_{12}^F$ and the boundary condition $g\left(Y_{11}^F\right) = \frac{Y_{11}^F D_{11}}{r-\mu}$:

$$\kappa_1 = \kappa_3 + \varepsilon_1, \quad (9.71)$$
$$\kappa_3 = \left(Y_{11}^F\right)^{-\beta_1^*}\left(\frac{Y_{11}^F D_{11}}{r-\mu} - \frac{\lambda Y_{11}^F D_{12}}{(r+\lambda-\mu)(r-\mu)}\right) - \left(Y_{11}^F\right)^{\beta_2^*-\beta_1^*}\varepsilon_2, \quad (9.72)$$
$$\kappa_4 = \varepsilon_2. \quad (9.73)$$

Combining equations (9.58), (9.59), (9.64), and (9.70) gives equation (9.29), in which

$$\phi_1 = v_1 + \kappa_3 + \varepsilon_1, \quad (9.74)$$
$$\phi_2 = v_1 + \kappa_3, \quad (9.75)$$
$$\phi_3 = \kappa_4 = \varepsilon_2. \quad (9.76)$$

Lemma 9.5 in Appendix B states that $\phi_1 > 0$ and $\phi_3 > 0$.

A.3   WAITING CURVE

The following Bellman equation must hold for the waiting curve

$$rW\left(Y\right) = \lim_{dt\downarrow 0}\frac{1}{dt}E\left[dW\left(Y\right)\right]. \quad (9.77)$$

Itô's lemma gives

$$E\left[dW\left(Y\right)\right] = \left(1 - \lambda dt\right)\left(\frac{\partial W\left(Y\right)}{\partial Y}\mu Y dt + \frac{\partial^2 W\left(Y\right)}{\partial Y^2}\frac{1}{2}\sigma^2 Y^2 dt\right)$$
$$+ \lambda dt\left(\Phi_{22}\left(Y\right) - W\left(Y\right)\right) + o\left(dt\right). \quad (9.78)$$

Substitution of (9.78) in (9.77) gives

$$rW\left(Y\right) = \frac{\partial W\left(Y\right)}{\partial Y}\mu Y + \frac{\partial^2 W\left(Y\right)}{\partial Y^2}\frac{1}{2}\sigma^2 Y^2 + \lambda\left(\Phi_{22}\left(Y\right) - W\left(Y\right)\right).$$
$$(9.79)$$

Rewriting gives

$$\frac{1}{2}\sigma^2 Y^2 \frac{\partial^2 W(Y)}{\partial Y^2} + \mu Y \frac{\partial W(Y)}{\partial Y} - (r+\lambda) W(Y) + \lambda \Phi_{22}(Y) = 0.$$

$$(9.80)$$

Using equation (9.6) and the boundary condition for $Y = 0$ and ruling out speculative bubbles,

$$W(0) = 0, \qquad (9.81)$$

$$\lim_{Y \to \infty} \frac{W(Y)}{Y} = \frac{\lambda D_{22}}{(r+\lambda-\mu)(r-\mu)}, \qquad (9.82)$$

gives

$$W(Y) = \begin{cases} \eta_1 Y^{\beta_1^*} + A_{22} Y^{\beta_1} & \text{if } Y < Y_{22}^F, \\ \eta_2 Y^{\beta_2^*} + \frac{\lambda Y D_{22}}{(r+\lambda-\mu)(r-\mu)} - \frac{\lambda I}{r+\lambda} & \text{if } Y \geq Y_{22}^F. \end{cases} \qquad (9.83)$$

The constants η_1 and η_2 are found by solving the continuity and differentiability conditions for W at $Y = Y_{22}^F$:

$$\eta_1 = \frac{\left(Y_{22}^F\right)^{-\beta_1^*} I\left(r(r-\mu)\beta_2^* + (r-\mu\beta_1)\lambda\beta_2^* - (r-\mu)(r+\lambda)\beta_1\right)}{(r+\lambda)(r+\lambda-\mu)(\beta_1-1)\left(\beta_1^*-\beta_2^*\right)}, \qquad (9.84)$$

$$\eta_2 = \frac{\left(Y_{22}^F\right)^{-\beta_2^*} I\left(r(r-\mu)\beta_1^* + (r-\mu\beta_1)\lambda\beta_1^* - (r-\mu)(r+\lambda)\beta_1\right)}{(r+\lambda)(r+\lambda-\mu)(\beta_1-1)\left(\beta_1^*-\beta_2^*\right)}. \qquad (9.85)$$

A direct result of Lemma 9.1 is that $\eta_1 < 0$ and $\eta_2 > 0$.

B. LEMMAS AND PROOFS

LEMMA 9.1 *The following two inequalities hold:*

$$r(r-\mu)\beta_2^* + (r-\mu\beta_1)\lambda\beta_2^* - (r-\mu)(r+\lambda)\beta_1 < 0, \quad (9.86)$$
$$r(r-\mu)\beta_1^* + (r-\mu\beta_1)\lambda\beta_1^* - (r-\mu)(r+\lambda)\beta_1 > 0. \quad (9.87)$$

PROOF OF LEMMA 9.1 *The assumption $\mu \in (0,r)$ implies that (see Proof of Proposition 7.5):*

$$1 \leq \beta_1 \leq \frac{r}{\mu} \qquad (9.88)$$

Equation (9.86) holds due to equation (9.88) and the fact that $\beta_2^* < 0$.
We know that $\beta_1^* \geq \beta_1$, where the equality sign only holds for $\sigma \to \infty$ for which we have $\beta_1^* = \beta_1 = 1$. Write $\beta_1^* = \xi\beta_1$ and substitute in (9.87):

$$r(r-\mu)\xi\beta_1 + (r-\mu\beta_1)\lambda\xi\beta_1 - (r+\lambda)(r-\mu)\beta_1$$
$$= r(r+\lambda-\mu)\beta_1(\xi-1) - \mu\lambda\beta_1(\xi\beta_1-1)$$
$$= \Xi(\xi). \qquad (9.89)$$

Then $\Xi\left(1\right) = 0$ ($\xi = 1$ implies that $\beta_1 = \beta_1^*$ and therefore $\beta_1 = 1$) and

$$\frac{d\Xi\left(\xi\right)}{d\xi} = \beta_1\left(r\left(r+\lambda\right) - \mu\left(r+\lambda\beta_1\right)\right) > 0, \tag{9.90}$$

if and only if

$$r\left(r+\lambda\right) - \mu\left(r+\lambda\beta_1\right) > 0. \tag{9.91}$$

Equation (9.91) holds since

$$\beta_1 < \frac{r}{\mu} + \frac{r\left(r-\mu\right)}{\mu\lambda}. \tag{9.92}$$

Therefore equation (9.87) holds. □

LEMMA 9.2 The constants δ_1, δ_2, and δ_3 have the following signs: $\delta_1 < 0$, $\delta_2 > 0$, and $\delta_3 > 0$.

PROOF OF LEMMA 9.2 The signs of δ_2 and δ_3 follow immediately from equations (9.47) and (9.48). Define the following functions:

$$Z\left(Y\right) = Y^{\beta_2^* - \beta_1^*}\frac{\left(1-\beta_2^*\right)\gamma_2}{\beta_1^* - 1} + Y^{-\beta_1^*}\frac{rI}{\left(\beta_1^* - 1\right)\left(r+\lambda\right)}, \tag{9.93}$$

$$V\left(Y\right) = Z\left(Y\right) + \left(Y_{12}^F\right)^{\beta_2^* - \beta_1^*}\frac{\beta_2^*\gamma_2}{\beta_1^*} - \left(Y_{12}^F\right)^{1-\beta_1^*}\frac{D_{21}}{\beta_1^*\left(r+\lambda-\mu\right)}. \tag{9.94}$$

The first derivative of Z is negative and it can be checked that $V\left(Y_{12}^F\right) = 0$. Therefore, because $Y_{11}^F > Y_{12}^F$ it holds that $\delta_1 = V\left(Y_{11}^F\right) < V\left(Y_{12}^F\right) = 0$. □

LEMMA 9.3 It holds that

$$\sqrt{2r\sigma^2 + \left(\mu - \frac{1}{2}\sigma^2\right)^2} \geq \mu + \frac{1}{2}\sigma^2. \tag{9.95}$$

PROOF OF LEMMA 9.3 Squaring both sides of (9.95) gives

$$2r\sigma^2 + \left(\mu - \frac{1}{2}\sigma^2\right)^2 \geq \left(\mu + \frac{1}{2}\sigma^2\right)^2. \tag{9.96}$$

Rewriting of (9.96) yields

$$2\left(r-\mu\right)\sigma^2 \geq 0. \tag{9.97}$$

Therefore the lemma holds since we assumed that $r > \mu$. □

LEMMA 9.4 *The constants ε_1 and ε_2 are both positive.*

PROOF OF LEMMA 9.4 *From equation (9.56) it follows that $\varepsilon_1 > 0$. The lemma holds whenever the following statement is true:*

$$(r - \mu)(\beta_1 - \beta_1^*) + \lambda(\beta_1 - 1) > 0. \tag{9.98}$$

In order to prove that equation (9.98) holds, define the following function

$$\Omega(\lambda) = (r - \mu)(\beta_1 - \beta_1^*(\lambda)) + \lambda(\beta_1 - 1). \tag{9.99}$$

For $\lambda = 0$ we have that $\beta_1 = \beta_1^*$, so that $\Omega(0) = 0$. The second derivative of Ω is equal to

$$\frac{\partial^2 \Omega(\lambda)}{\partial \lambda^2} = \frac{(r - \mu)\sigma^2}{\left(2(r + \lambda)\sigma^2 + \left(\mu - \frac{1}{2}\sigma^2\right)^2\right)^{\frac{3}{2}}} > 0. \tag{9.100}$$

Thus the lemma is proved if it holds that

$$\left.\frac{\partial \Omega(\lambda)}{\partial \lambda}\right|_{\lambda=0} \geq 0. \tag{9.101}$$

The first derivative of Ω at $\lambda = 0$ equals

$$\left.\frac{\partial \Omega(\lambda)}{\partial \lambda}\right|_{\lambda=0} = \frac{4\mu^2 + 4r\sigma^2 + \sigma^4 - (4\mu + 2\sigma^2)\sqrt{2r\sigma^2 + \left(\mu - \frac{1}{2}\sigma^2\right)^2}}{4\sigma^2\sqrt{2r\sigma^2 + \left(\mu - \frac{1}{2}\sigma^2\right)^2}}. \tag{9.102}$$

Define

$$\eta(r) = 4\mu^2 + 4r\sigma^2 + \sigma^4 - (4\mu + 2\sigma^2)\sqrt{2r\sigma^2 + \left(\mu - \frac{1}{2}\sigma^2\right)^2}. \tag{9.103}$$

Then $\eta(\mu) = 0$ and with Lemma 9.4 we have

$$\frac{\partial \eta(r)}{\partial r} = \frac{4\sigma^2\left(\sqrt{2r\sigma^2 + \left(\mu - \frac{1}{2}\sigma^2\right)^2} - \left(\mu + \frac{1}{2}\sigma^2\right)\right)}{\sqrt{2r\sigma^2 + \left(\mu - \frac{1}{2}\sigma^2\right)^2}} \geq 0. \tag{9.104}$$

This implies that for $r \geq \mu$ equation (9.101) holds and thereby the lemma is proved. $\qquad\square$

LEMMA 9.5 *The constants ϕ_1 and ϕ_3 have the following signs: $\phi_1 > 0$ and $\phi_3 > 0$.*

PROOF OF LEMMA 9.5 *We start with ϕ_3. From Lemma 9.5 we know that $\varepsilon_2 > 0$. Therefore $\phi_3 > 0$.*
  Define

$$E_1(Y) = \frac{Y^{1-\beta_1^*}\left((r-\mu)(\beta_1-\beta_2^*)+\lambda(\beta_1-1)\right)(D_{10}-D_{12})}{(r+\lambda-\mu)(r-\mu)(\beta_1^*-\beta_2^*)}, \quad (9.105)$$

$$E_2(Y) = \frac{Y^{1-\beta_2^*}\left((r-\mu)(\beta_1-\beta_1^*)+\lambda(\beta_1-1)\right)(D_{10}-D_{12})}{(r+\lambda-\mu)(r-\mu)(\beta_1^*-\beta_2^*)}. \quad (9.106)$$

Then $E_1\left(Y_{12}^F\right) = \varepsilon_1$ and $E_2\left(Y_{12}^F\right) = \varepsilon_2$, further it holds that $E_1\left(Y_{11}^F\right) < E_1\left(Y_{12}^F\right)$ and $E_2\left(Y_{11}^F\right) > E_2\left(Y_{12}^F\right)$. Further define

$$K_3(Y) = \left(Y_{11}^F\right)^{-\beta_1^*}\left(\frac{Y_{11}^F D_{11}}{r-\mu} - \frac{\lambda Y_{11}^F D_{12}}{(r+\lambda-\mu)(r-\mu)}\right) - \left(Y_{11}^F\right)^{\beta_2^*-\beta_1^*} E_2(Y),$$
$$(9.107)$$

then it follows after some tedious calculations that

$$\begin{aligned}
\phi_1 &= v_1 + \kappa_3 + \varepsilon_1 \\
&= v_1 + K_3\left(Y_{12}^F\right) + E_1\left(Y_{12}^F\right) \\
&> v_1 + K_3\left(Y_{11}^F\right) + E_1\left(Y_{11}^F\right) \\
&= \frac{\left(Y_{11}^F\right)^{1-\beta_1^*}(D_{11}-D_{12})}{(r+\lambda-\mu)(r-\mu)} > 0. \quad (9.108)
\end{aligned}$$

Thus ϕ_1 is positive. □

PROOF OF PROPOSITION 9.1 *It is easy to verify that equation (9.24) does not have a root if $\lambda \geq \lambda_1^*$. Assertion 2 can be concluded by taking a closer look at equations (9.24), (9.20), (9.47), and (9.41). The closer λ comes to λ_1^* the smaller the negative term in (9.24) becomes in absolute terms. This implies that Y_{11}^F becomes larger.* □

PROOF OF PROPOSITION 9.2 *From Proposition 9.1 we know that Y_{11}^F does not exist for $\lambda \geq \lambda_1^*$ and therefore ϕ_2 does not make sense for $\lambda \geq \lambda_1^*$. First we prove that $\phi_2 \leq 0$ if equation (9.30) holds and $\lambda < \lambda_1^*$. According to (9.65), (9.72), and (9.75) it is sufficient to prove that*

$$\frac{D_{11}}{r-\mu} - \frac{\lambda D_{12}}{(r+\lambda-\mu)(r-\mu)} - \frac{D_{10}}{r+\lambda-\mu} \leq 0. \quad (9.109)$$

Equation (9.109) holds if

$$\lambda \leq \frac{(r-\mu)(D_{10}-D_{11})}{D_{11}-D_{12}}. \quad (9.110)$$

Using equation (9.30) it is not hard to show that

$$\frac{(r - \mu)(D_{10} - D_{11})}{D_{11} - D_{12}} > \lambda_1^*. \tag{9.111}$$

Therefore equation (9.109) holds and ϕ_2 is non-positive.

Let us show that ϕ_2 can be negative when (9.30) does not hold. Set $\lambda = 0$, then $\beta_1 = \beta_1^*$ and $\varepsilon_2 = 0$ so that

$$\phi_2 = \left(Y_{11}^F\right)^{1-\beta_1} \frac{D_{11} - D_{10}}{r - \mu} < 0. \tag{9.112}$$

Next we argue that ϕ_2 can be positive when equation (9.30) does not hold. Define the following function

$$\mathcal{F}_2(Y) = Y^{1-\beta_1^*} \frac{(r - \mu)(D_{11} - D_{10}) - \lambda(D_{12} - D_{11})}{(r + \lambda - \mu)(r - \mu)} - Y^{\beta_2^* - \beta_1^*} \varepsilon_2. \tag{9.113}$$

Thus $\mathcal{F}_2\left(Y_{11}^F\right) = \phi_2$. When equation (9.30) does not hold, the first term in equation (9.113) is positive. When λ approaches λ_1^* we know from Proposition 9.1 that Y_{11}^F approaches infinity. Taking a closer look at equation (9.113) we see that the second term goes faster to zero than the first term. Thus for λ close enough to λ_1^* we have that ϕ_2 is positive. □

PROOF OF PROPOSITION 9.3 *This proposition is easily verified by taking a closer look at equations (9.23), (9.32), (9.40), (9.41), (9.84), and (9.85).* □

PROOF OF PROPOSITION 9.4 *Define the function $\mathcal{L}$ as follows*

$$\mathcal{L}(Y) = L_{11}(Y) - F_{11}(Y). \tag{9.114}$$

The functions L_{11} and F_{11} are continuous. Further it holds that $\mathcal{L}(0) = -I$ and $\mathcal{L}\left(Y_{11}^F\right) = 0$. Therefore it is sufficient to prove that

$$\left.\frac{\partial \mathcal{L}(Y)}{\partial Y}\right|_{Y=Y_{11}^F} < 0. \tag{9.115}$$

Substitution of equations (9.23) and (9.29) in (9.115) gives for $Y \in \left[Y_{12}^F, Y_{11}^F\right]$:

$$\begin{aligned} \mathcal{L}(Y) = {} & (\phi_2 - \delta_2) Y^{\beta_1^*} + (\phi_3 - \delta_3) Y^{\beta_2^*} \\ & + \frac{(r - \mu) D_{10} - \lambda(D_{21} - D_{12})}{(r + \lambda - \mu)(r - \mu)} Y - \frac{rI}{r + \lambda}. \end{aligned} \tag{9.116}$$

Thus

$$\frac{\partial \mathcal{L}(Y)}{\partial Y} = \beta_1^* (\phi_2 - \delta_2) Y^{\beta_1^* - 1} + \beta_2^* (\phi_3 - \delta_3) Y^{\beta_2^* - 1}$$
$$+ \frac{(r - \mu) D_{10} - \lambda (D_{21} - D_{12})}{(r + \lambda - \mu)(r - \mu)}. \qquad (9.117)$$

From $\mathcal{L}\left(Y_{11}^F\right) = 0$ *we obtain*

$$\frac{Y_{11}^F}{\beta_1^*} \frac{\partial \mathcal{L}(Y)}{\partial Y}\bigg|_{Y = Y_{11}^F} = \frac{\beta_2^* - \beta_1^*}{\beta_1^*} (\phi_3 - \delta_3)\left(Y_{11}^F\right)^{\beta_2^*} + \frac{rI}{r + \lambda}$$
$$+ \frac{1 - \beta_1^*}{\beta_1^*} \frac{(r - \mu) D_{10} - \lambda (D_{21} - D_{12})}{(r + \lambda - \mu)(r - \mu)} Y_{11}^F. \qquad (9.118)$$

Subtracting $\frac{1}{\beta_1^*}$ *times equation (9.24) from equation (9.118) gives*

$$\frac{Y_{11}^F}{\beta_1^*} \frac{\partial \mathcal{L}(Y)}{\partial Y}\bigg|_{Y = Y_{11}^F} = \frac{\beta_2^* - \beta_1^*}{\beta_1^*} \phi_3 \left(Y_{11}^F\right)^{\beta_2^*} + \frac{1 - \beta_1^*}{\beta_1^*}$$
$$\times \frac{\lambda (D_{12} - D_{11}) - (r - \mu)(D_{11} - D_{10})}{(r + \lambda - \mu)(r - \mu)} Y_{11}^F. \qquad (9.119)$$

From the proof of Proposition 9.2 we know that

$$\lambda \le \frac{(r - \mu)(D_{10} - D_{11})}{D_{11} - D_{12}}. \qquad (9.120)$$

Equations (9.119) and (9.120) together with $\phi_3 > 0$ *imply equation (9.115).* □

PROOF OF PROPOSITION 9.5 *Define* $P(Y) = \Pr\left(T_S < T \mid Y(0) = Y\right)$. *Then for* $Y < S$ *the function* P *must satisfy the following Bellman equation*

$$0 = -\lambda P(Y) + \frac{\partial P(Y)}{\partial Y} \mu Y + \frac{\partial^2 P(Y)}{\partial Y^2} \frac{1}{2} \sigma^2 Y^2. \qquad (9.121)$$

Since $P(0) = 0$ *and* $P(S) = 1$ *the solution of (9.121) equals (9.36).* □

PROOF OF PROPOSITION 9.6 *Taking a closer look at equations (9.19) and (9.28) (for Y large) we see that there exists a crossing point of* L_{12} *and* F_{12} *if*

$$\frac{Y D_{10}}{r + \lambda - \mu} + \frac{\lambda}{r + \lambda - \mu} \frac{Y D_{12}}{r - \mu} > \frac{\lambda}{r + \lambda - \mu} \frac{Y D_{21}}{r - \mu}. \qquad (9.122)$$

Rewriting (9.122) gives (9.33). □

PROOF OF PROPOSITION 9.7 *This proof follows the same lines as the proof of Proposition 9.6, but then with equations (9.28) and (9.32).* □

References

Akerlof, G. A. (1970). The market for lemons: Qualitative uncertainty and the market mechanism. *The Quarterly Journal of Economics*, 84:488–500.

Amram, M. and Kulatilaka, N. (1998). *Real Options: Managing Strategic Investment in an Uncertain World*. Harvard Business School Press, Watertown, Massachusetts, United States of America.

Balcer, Y. and Lippman, S. A. (1984). Technological expectations and adoption of improved technology. *Journal of Economic Theory*, 34: 292–318.

Baldursson, F. M. (1998). Irreversible investment under uncertainty in oligopoly. *Journal of Economic Dynamics and Control*, 22:627–644.

Baldwin, C. Y. (1982). Optimal sequential investment when capital is not readily reversible. *The Journal of Finance*, 37:763–782.

Brealey, R. A. and Myers, S. C. (1991). *Principles of Corporate Finance*, fourth edition. McGraw-Hill, New York, New York, United States of America.

Bridges, E., Coughlan, A. T., and Kalish, S. (1991). New technology adoption in an innovative marketplace: Micro- and macro-level decision making models. *International Journal of Forecasting*, 7:257–270.

Brynjolfsson, E., Malone, T., and Gurbaxani, V. (1991). Does information technology lead to smaller firms? Technical Report 123, Center of Coordination Science, MIT, Cambridge, Massachusetts, United States of America.

Dixit, A. K. (1991). A simplified treatment of the theory of optimal regulation of brownian motion. *Journal of Economic Dynamics and Control*, 15:657–673.

Dixit, A. K. (1993). *The Art of Smooth Pasting*. Harwood Academic Publishers, Chur, Switzerland.

Dixit, A. K. and Pindyck, R. S. (1996). *Investment Under Uncertainty*, second printing. Princeton University Press, Princeton, New Jersey, United States of America.

Doraszelski, U. (2001). The net present value method versus the option value of waiting: A note on farzin, huisman and kort (1998). *Journal of Economic Dynamics and Control*, 25:1109–1115.

Dosi, C. and Moretto, M. (1997). Pollution accumulation and firm incentives to accelerate technological change under uncertain private benefits. *Environmental and Resource Economics*, 10:285–300.

Dutta, P. K., Lach, S., and Rustichini, A. (1995). Better late than early: Vertical differentiation in the adoption of a new technology. *Journal of Economics and Management Strategy*, 4:563–589.

Ekboir, J. M. (1997). Technical change and irreversible investment under risk. *Agricultural Economics*, 16:55–65.

Farzin, Y. H., Huisman, K. J. M., and Kort, P. M. (1998). Optimal timing of technology adoption. *Journal of Economic Dynamics and Control*, 22:779–799.

Fudenberg, D. and Tirole, J. (1985). Preemption and rent equalization in the adoption of new technology. *The Review of Economic Studies*, 52:383–401.

Fudenberg, D. and Tirole, J. (1991). *Game Theory*. The MIT Press, Cambridge, Massachusetts, United States of America.

Gaimon, C. (1989). Dynamic game results of the acquisition of new technology. *Operations Research*, 37:410–425.

Götz, G. (2000). Strategic timing of adoption of new technologies under uncertainty: A note. *International Journal of Industrial Organization*, 18:369–379.

Grenadier, S. R. (1996). The strategic exercise of options: Development cascades and overbuilding in real estate markets. *The Journal of Finance*, 51:1653–1679.

Grenadier, S. R. and Weiss, A. M. (1997). Investment in technological innovations: An option pricing approach. *Journal of Financial Economics*, 44:397–416.

Hendricks, K. (1992). Reputations in the adoption of a new technology. *International Journal of Industrial Organization*, 10:663–677.

Hendricks, K., Weiss, A., and Wilson, C. (1988). The war of attrition in continuous time with complete information. *International Economic Review*, 29:663–680.

Hendricks, K. and Wilson, C. (1992). Equilibrium in preemption games with complete information. In Majumdar, M., editor, *Essays in Honour of David Gale*, pages 123–147. St. Martin's Press, New York, New York, United States of America.

Hoppe, H. C. (2000). Second-mover advantages in the strategic adoption of new technology under uncertainty. *International Journal of Industrial Organization*, 18:315–338.

Huisman, K. J. M. (1996). Optimal timing of technology adoption. Master's thesis, Tilburg University, Department of Econometrics, Tilburg, The Netherlands.

Huisman, K. J. M. and Kort, P. M. (1998a). A further analysis on strategic timing of adoption of new technologies under uncertainty. CentER Discussion Paper 9803, Tilburg University, CentER, Tilburg, The Netherlands.

Huisman, K. J. M. and Kort, P. M. (1998b). Strategic investment in technological innovations. CentER Discussion Paper 98114, Tilburg University, CentER, Tilburg, The Netherlands.

Huisman, K. J. M. and Kort, P. M. (1999). Effects of strategic interactions on the option value of waiting. CentER Discussion Paper 9992, Tilburg University, CentER, Tilburg, The Netherlands.

Huisman, K. J. M. and Kort, P. M. (2000). Strategic technology adoption taking into account future technological improvements: A real options approach. CentER Discussion Paper 2000-52, Tilburg University, CentER, Tilburg, The Netherlands.

Huisman, K. J. M. and Kort, P. M. (2001). Strategic technology investment under uncertainty. *OR Spektrum*. to appear.

Huisman, K. J. M. and Nielsen, M. J. (2001). Asymmetric firms and irreversible investments. Working Paper, Tilburg University, Department of Econometrics and CentER, Tilburg, The Netherlands, and University of Copenhagen, Institute of Economics, Copenhagen, Denmark.

Hull, J. C. (1993). *Options, Futures, and Other Derivative Securities*, second edition. Prentice-Hall, London, United Kingdom.

Karatzas, I. and Shreve, S. E. (1991). *Brownian Motion and Stochastic Calculus*, second edition. Springer-Verlag, New York, New York, United States of America.

Kriebel, C. H. (1989). Understanding the strategic investment in information technology. In Laudon, K. C. and Turner, J. A., editors, *Information Technology and Management Strategy*. Prentice-Hall, Englewood Cliffs, New Jersey, United States of America.

Kulatilaka, N. and Perotti, E. C. (1998). Strategic growth options. *Management Science*, 44:1021–1031.

Lambrecht, B. and Perraudin, W. (1999). Real options and preemption under incomplete information. Working Paper, University of London, The Institute for Financial Research, Birkbeck College, London, United Kingdom.

Lander, D. M. and Pinches, G. E. (1998). Challenges to the practical implementation of modeling and valuing real options. *The Quarterly Review of Economics and Finance*, 38:537–567.

McDonald, R. and Siegel, D. R. (1986). The value of waiting to invest. *The Quarterly Journal of Economics*, 101:707–727.

Merton, R. C. (1992). *Continuous Time Finance*, revised edition. Blackwell Publishers, Oxford, United Kingdom.

Nair, S. K. (1995). Modeling strategic investment decisions under sequential technological change. *Management Science*, 41:282–297.

Nielsen, M. J. (1999). Competition and irreversible investments. Working Paper, University of Copenhagen, Institute of Economics, Copenhagen, Denmark.

Pawlina, G. and Kort, P. M. (2001). Investment under uncertainty and policy change. CentER Discussion Paper 2001-05, Tilburg University, CentER, Tilburg, The Netherlands.

Pennings, H. P. G. (1998). *Real Options and Managerial Decision Making*. PhD thesis, Erasmus University, Rotterdam Institute for Business Economic Studies, Rotterdam, The Netherlands.

Pennings, H. P. G. and Sleuwaegen, L. (1998). The choice and timing of foreign market entry under uncertainty. Onderzoeksrapport 9826, Katholieke Universiteit Leuven, Departement Toegepaste Economische Wetenschappen, Leuven, Belgium.

Purvis, A., Boggess, W. G., Moss, C. B., and Holt, J. (1995). Technology adoption decisions under irreversibility and uncertainty: An *ex ante* approach. *American Journal of Agricultural Economics*, 77:541–551.

Rajagopalan, S. (1999). Adoption timing of new equipment with another innovation anticipated. *IEEE Transactions on Engineering Management*, 46:14–25.

Rajagopalan, S., Singh, M. R., and Morton, T. E. (1998). Capacity expansion and replacement in growing markets with uncertain technological breakthroughs. *Management Science*, 44:12–30.

Reinganum, J. F. (1981). On the diffusion of new technology: A game theoretic approach. *The Review of Economic Studies*, 48:395–405.

Rozendaal, S. (1998). Almaar krimpende chips. *Elsevier*, 4:86–90.

Sarkar, S. (2000). On the investment-uncertainty relationship in a real options model. *Journal of Economic Dynamics and Control*, 24:219–225.

Simon, L. K. (1987a). Basic timing games. Working Paper 8745, University of California at Berkeley, Department of Economics, Berkeley, California, United States of America.

Simon, L. K. (1987b). A multistage duel in continuous time. Working Paper 8757, University of California at Berkeley, Department of Economics, Berkeley, California, United States of America.

Simon, L. K. and Stinchcombe, M. B. (1989). Extensive form games in continuous time: Pure strategies. *Econometrica*, 57:1171–1214.

Smets, F. (1991). Exporting versus FDI: The effect of uncertainty, irreversibilities and strategic interactions. Working Paper, Yale University, New Haven, Connecticut, United States of America.

Smit, J. T. J. (1996). *Growth Options and Strategy Analysis*. PhD thesis, Universiteit van Amsterdam, Amsterdam, The Netherlands.

Smit, J. T. J. (1997). Investment analysis of offshore concessions in the netherlands. *Financial Management*, 26:5–17.

Smit, J. T. J. and Ankum, L. A. (1993). A real options and game-theoretic approach to corporate investment strategy under competition. *Financial Management*, 22:241–250.

Somma, E. (1999). The effect of incomplete information about future technological opportunities on pre-emption. *International Journal of Industrial Organization*, 17:765–799.

Stenbacka, R. and Tombak, M. M. (1994). Strategic timing of adoption of new technologies under uncertainty. *International Journal of Industrial Organization*, 12:387–411.

Thijssen, J. J. J., Huisman, K. J. M., and Kort, P. M. (2001a). Strategic investment under uncertainty and information spillovers. Working Paper, Tilburg University, Department of Econometrics and CentER, Tilburg, The Netherlands.

Thijssen, J. J. J., van Damme, E. E. C., Huisman, K. J. M., and Kort, P. M. (2001b). Investment under vanishing uncertainty due to information arriving over time. CentER Discussion Paper 2001-14, Tilburg University, CentER, Tilburg, The Netherlands.

Tirole, J. (1988). *The Theory of Industrial Organization*. The MIT Press, Cambridge, Massachusetts, United States of America.

Torvund, P. A. (1999). *Strategic Timing Behaviour and Related Game Theoretical Topics*. PhD thesis, University of Oslo, Department of Mathematics, Oslo, Norway.

Trigeorgis, L. (1995). *Real Options in Capital Investment: Models, Strategies, and Applications*. Praeger Publisher, Westport, Connecticut, United States of America.

Trigeorgis, L. (1996). *Real Options: Managerial Flexibility and Strategy in Resource Allocation*. The MIT Press, Cambridge, Massachusetts, United States of America.

van den Goorbergh, R. W. J., Huisman, K. J. M., and Kort, P. M. (2001). Investment under uncertainty and risk aversion in incomplete markets. Working Paper, Tilburg University, Department of Finance and CentER, Tilburg, The Netherlands.

Weeds, H. (1999). Strategic delay in a real options model of r & d competition. Working Paper, University of Cambridge, Fitzwilliam College, Cambridge, United Kingdom.

Yorukoglu, M. (1998). The information technology productivity paradox. *Review of Economic Dynamics*, 1:551–592.

Author Index

Subject Index

Bold page numbers refer to definitions.

About the Author

Dr. Kuno J.M. Huisman graduated in Econometrics from Tilburg University in 1996 with a specialization in Management Science. After that, he carried out his Ph.D. research at the Department of Econometrics and CentER at the same university. As of September 2000 he works as consultant at the Centre for Quantitative Methods in Eindhoven, The Netherlands.

THEORY AND DECISION LIBRARY

SERIES C: GAME THEORY, MATHEMATICAL PROGRAMMING AND OPERATIONS RESEARCH
Editor: S.H. Tijs, *University of Tilburg, The Netherlands*

24. J. Suijs: *Cooperative Decision-Making under Risk.* 1999   ISBN 0-7923-8660-4
25. J. Rosenmüller: *Game Theory: Stochastics, Information, Strategies and Cooperation.* 2000   ISBN 0-7923-8673-6
26. J.M. Bilbao: *Cooperative Games on Combinatorial Structures.* 2000
    ISBN 0-7923-7782-6
27. M. Slikker and A. van den Nouweland: *Social and Economic Networks in Cooperative Game Theory.* 2000   ISBN 0-7923-7226-3
28. K.J.M. Huisman: *Technology Investment: A Game Theoretic Real Options Approach.* 2001   ISBN 0-7923-7487-8
29. A. Perea: *Rationality in Extensive Form Games.* 2001   ISBN 0-7923-7540-8

KLUWER ACADEMIC PUBLISHERS – DORDRECHT / BOSTON / LONDON